101 SOLVED MECHANICAL ENGINEERING PROBLEMS

Michael R. Lindeburg, P.E.

PROFESSIONAL PUBLICATIONS, INC.
Belmont, CA 94002

In the ENGINEERING LICENSING EXAM AND REFERENCE SERIES

Engineer-In-Training Reference Manual
EIT Review Manual
Engineering Fundamentals Quick Reference Cards
Engineer-In-Training Sample Examinations
Mini-Exams for the E-I-T Exam
1001 Solved Engineering Fundamentals Problems
Fundamentals of Engineering Exam Study Guide
Diagnostic F.E. Exam for the Macintosh
Fundamentals of Engineering Video Series: Thermodynamics
Civil Engineering Reference Manual
Civil Engineering Quick Reference Cards
Civil Engineering Sample Examination
Civil Engineering Review Course on Cassettes
101 Solved Civil Engineering Problems
Seismic Design of Building Structures
Seismic Design Fast
345 Solved Seismic Design Problems
Timber Design for the Civil P.E. Exam
Fundamentals of Reinforced Masonry Design
246 Solved Structural Engineering Problems
Mechanical Engineering Reference Manual
Mechanical Engineering Quick Reference Cards
Mechanical Engineering Sample Examination
101 Solved Mechanical Engineering Problems
Mechanical Engineering Review Course on Cassettes
Consolidated Gas Dynamics Tables
Fire and Explosion Protection Systems
Electrical Engineering Reference Manual
Electrical Engineering Quick Reference Cards
Electrical Engineering Sample Examination
Chemical Engineering Reference Manual
Chemical Engineering Quick Reference Cards
Chemical Engineering Practice Exam Set
Land Surveyor Reference Manual
Land Surveyor-In-Training Sample Examination
1001 Solved Surveying Fundamentals Problems

In the GENERAL ENGINEERING and CAREER ADVANCEMENT SERIES

How to Become a Professional Engineer
Getting Started as a Consulting Engineer
The Expert Witness Handbook: A Guide for Engineers
Engineering Your Job Search
Engineering Your Start-Up
Intellectual Property Protection: A Guide for Engineers
High-Technology Degree Alternatives
Metric in Minutes
Engineering Economic Analysis
Engineering Law, Design Liability, and Professional Ethics
Engineering Unit Conversions

101 SOLVED MECHANICAL ENGINEERING PROBLEMS

Printed in the United States of America

ISBN: 0-912045-77-9

Professional Publications, Inc.
1250 Fifth Avenue, Belmont, CA 94002
(415) 593-9119

Current printing of this edition: 6

Revised and reprinted in 1995.

TABLE OF CONTENTS

PREFACE

These original problems are the most realistic, representative, and typical one-hour P.E. examination problems that I could write. I wrote the problems in this publication for you, to fill the gap between reviewing for the exam and actually taking the exam. Although these problems are original, they are structured around the same concepts that the mechanical engineering P.E. exam could test you on.

You will find that some of the problems in this book are very difficult, and others are more straightforward. To the extent that your backgrounds differ, the "difficult/easy" classification will be different for you than it is for another engineer. However, even within the areas that you feel competent, you will notice that some problems will take you an hour to solve, while others will only require 15 minutes. That's a good lesson to learn.

Some engineers will want to start at the beginning of this publication and work through all of the problems. That method works for some, but I would not recommend it as a primary review strategy. It is much better to review engineering theory from my *Mechanical Engineering Reference Manual,* your college textbooks, and other engineering handbooks.

How, then, should you use this book? That is a simple question for me to answer. After finishing your review of a subject, turn to the section in this book containing problems in that subject. See if you can pick out the 15-minute problems. See if you can outline the proper solution procedure ... before you look at the solutions. Then, work the problems without looking at the solutions I have prepared. Finally, check your work.

Of course, there is no guarantee that any problem similar to the types contained in this publication will appear on your exam or any exam in the future, but that is not the intent of this publication. Use *101 Solved Mechanical Engineering Problems* to sharpen your test-taking skills. I think you will be pleased with your results.

Michael R. Lindeburg, P.E.
Belmont, CA

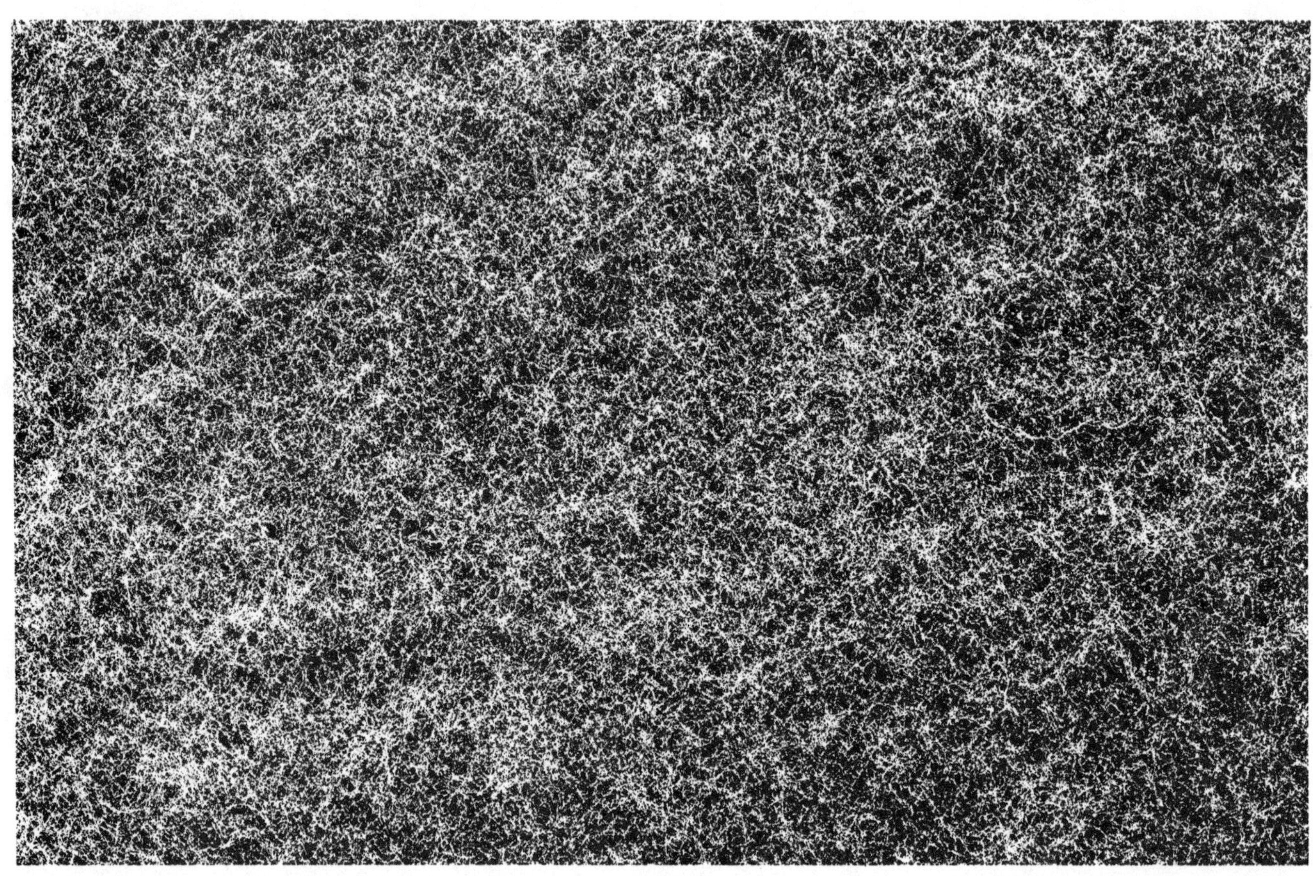

ACKNOWLEDGMENTS

With every book I write, I am reminded of how important it is to have a competent production team working together. Thank you, Jessica R. Whitney, for typesetting this publication, and Mia Laurence, for doing the proofreading. You both have great accuracy. Thank you, Jennifer Pasqual Thuillier, for illustrating the problems. And thank you, Lisa Rominger, for supervising the production and manufacturing processes. There were a lot of pieces to bring together.

I wish, though, that I could list the almost 200 mechanical engineers in my classes who reviewed these problems over a two-year period prior to publication. They were instrumental in knocking the rough edges off of preliminary versions of both problem statements and solutions.

Thanks, team! You're great.

Michael R. Lindeburg, P.E.

Notice To Examinees

Do not copy, memorize, or distribute problems from the Principles and Practice of Engineering (P&P) Examination. These acts are considered to be exam subversion.

The P&P examination is copyrighted by the National Council of Examiners for Engineering and Surveying. Copying and reproducing P&P exam problems for commercial purposes is a violation of federal copyright law. Reporting examination problems to other examinees invalidates the examination process and threatens the health and welfare of the public.

1 ENGINEERING ECONOMIC ANALYSIS

ECONOMICS–1

The Acme Manufacturing Company produces a product from polystyrene. Two mixing processes, batch and continuous, are available.

	batch	continuous
initial cost	\$25,000	\$50,000
lifetime	5 years	5 years
maintenance (per year)	\$20,000	\$18,000
salvage value	\$3000	\$6000
capacity (units/year)	950,000	1,000,000

Acme uses straight-line depreciation, pays 48% of its net income as income tax, and has an after-tax minimum attractive rate of return of 15%. The company has an infinite market and can sell all of the products it produces at \$0.10 per unit.

Which manufacturing process should Acme invest in?

ECONOMICS–2

A corporation plans to buy a car for \$10,000 and depreciate it over 5 years using the double declining balance method. However, the company intends to sell the car after 3 years for an estimated \$5500. Annual maintenance is estimated at \$500. The corporation also intends to take advantage of a special $3\frac{1}{3}\%$ tax credit available for assets kept at least 3 years. The corporation uses an after-tax rate of return of 15% on all investment analyses. The corporation pays 40% of its net income in taxes. What is the present worth of this car acquisition?

ECONOMICS–3

For the past 3 years, Bravado Testing Engineers has been paying its inspector \$0.30 per mile when a personal car is used to travel to job sites. Such usage is 12,000 miles per year. As an alternative, Bravado is considering buying a company truck, which would eliminate the use of the inspector's personal car. The estimated details of the purchase are

purchase price	\$10,000
annual maintenance (includes gas)	\$1800
practical useful life	4 years
salvage value after 4 years	\$2000

Bravado will depreciate the truck over 3 years at the rate of 25% the first year, 38% the second, and 37% the third. The company's combined federal and state tax rate is 40%, and the after-tax minimum attractive rate of return is 15%. What should Bravado do?

ECONOMICS–4

Investors are evaluating an office building that has a purchase price of \$525,000. The investors expect to sell the building for \$700,000 after 7 years. During the time that the building is held, the building will be depreciated using straight-line depreciation, an estimated life of 15 years, and an (assumed) salvage value of zero. (The full purchase price will be depreciated.) The effective tax rate of the investors is 40%.

During ownership, the estimated expenses will be \$25,000 per year, which will be offset by rental income of \$45,000 for the first 3 years and \$75,000 for the next 4 years. What will be the after-tax rate of return on this investment?

ECONOMICS–5

A numerically controlled metal former originally costing \$1,300,000 has already been in use for 7 years. It has been depreciated each year on a straight line basis with a 10-year life and a salvage value of \$300,000. The current market value of the metal former is \$400,000, and this amount is not expected to change during the next 3 years. The current operating expenses amount to \$200,000 per year, and this amount is also not expected to change.

It has been proposed to replace the existing machine with a new metal former costing \$800,000. The initial operating expenses will be $40,000 during the first year, increasing each year by $35,000 during its estimated 10-year lifetime. The salvage value of the new metal former depends on the year in which it will be retired.

year retired	salvage value
1	$650,000
2	$550,000
3	$500,000
4	$450,000
5	$400,000
6	$350,000
7	$300,000
8	$250,000
9	$200,000
10	$150,000

The company intends to stay in business and will need a metal former for the foreseeable future. If a before-tax analysis is wanted, and the interest rate is 15%, should the existing metal former be replaced?

ECONOMICS–6

An entrance road into a large federal forest is needed. The road is expected to be in use for 30 years before major resurfacing or replacement is required. Four alternatives have been proposed. A minimum rate of return of 8% is required by government procurement regulations. Which alternative should be chosen? (All amounts are in dollars.)

		annual revenue		
alternative	initial cost	years 1–10	years 11–20	years 21–30
A	200,000	14,000	14,000	14,000
B	400,000	8000	20,000	8000
C	320,000	6000	6000	10,000
D	300,000	40,000	0	40,000

ECONOMICS–7

You have $25,000 available, and your accountant has provided you with four investment alternatives. Your after-tax minimum attractive rate of return is 12%. (The cash flows accompanying each alternative are after-tax.)

time	A	B	C	D
$t = 0$	−8000	−10,000	−8000	−14,000
$t = 1$	+3000	+ 2500	+4000	+ 2000
$t = 2$	+3000	+ 3000	+1600	+ 4000
$t = 3$	+3000	+ 3500	+1200	+ 6000
$t = 4$	+3000	+ 4000	+ 800	+ 8000

Alternatives A and B are independent, but alternative C is not. However, if alternative D is chosen, then alternative C must also be selected. If alternative A is chosen, then alternative C cannot be chosen. Which alternatives should you invest in?

ECONOMICS–8

A consortium of investors (combined income tax rate of 40%, minimum attractive rate of return of 25%) is considering the construction of a recreational water slide. Due to variations in weather, the number of days the slide can be open varies each year.

days of operation	% of years
80	15%
100	60%
120	25%

The entrance fee will be $5 per person per day. The daily attendance is expected to drop off the longer the slide is open during the year.

days of operation	attendance
1–80	500 people/day
81–100	450 people/day
101–120	350 people/day

The consortium has assumed the following values for the purpose of its evaluation.

initial cost	$375,000
lifetime	5 years
salvage value	$50,000
depreciation method	straight line
tax credit available	10%
maintenance (fixed)	$10,000 per year
labor costs	$300 each day slide is open

(a) What is the after-tax income or loss per year?

(b) Is the investment attractive over its lifetime?

(c) What is the after-tax payback period?

ECONOMICS–9

A fresh produce wholesaler is planning to have a larger produce-washing facility constructed. The construction cost as bid is $2,500,000 if paid in advance, but the wholesaler has contracted with its construction company to make three progress payments (to be adjusted for 4% annual inflation) as follows.

end of year 1	50% of bid cost
end of year 2	25% of bid cost
end of year 3	25% of bid cost

The wholesaler's minimum attractive rate of return is 12%. The wholesaler pays 37% federal income tax and 6% state income tax. The state tax is deductible on the federal tax return. The larger washing facility will be depreciated over 25 years (assuming a zero salvage value) using the straight-line method.

(a) What is the after-tax present worth of produce-related revenue required to pay for the expansion?

(b) What annual after-tax income is required to pay for the expansion in 25 years?

(c) Did the construction company make a good deal?

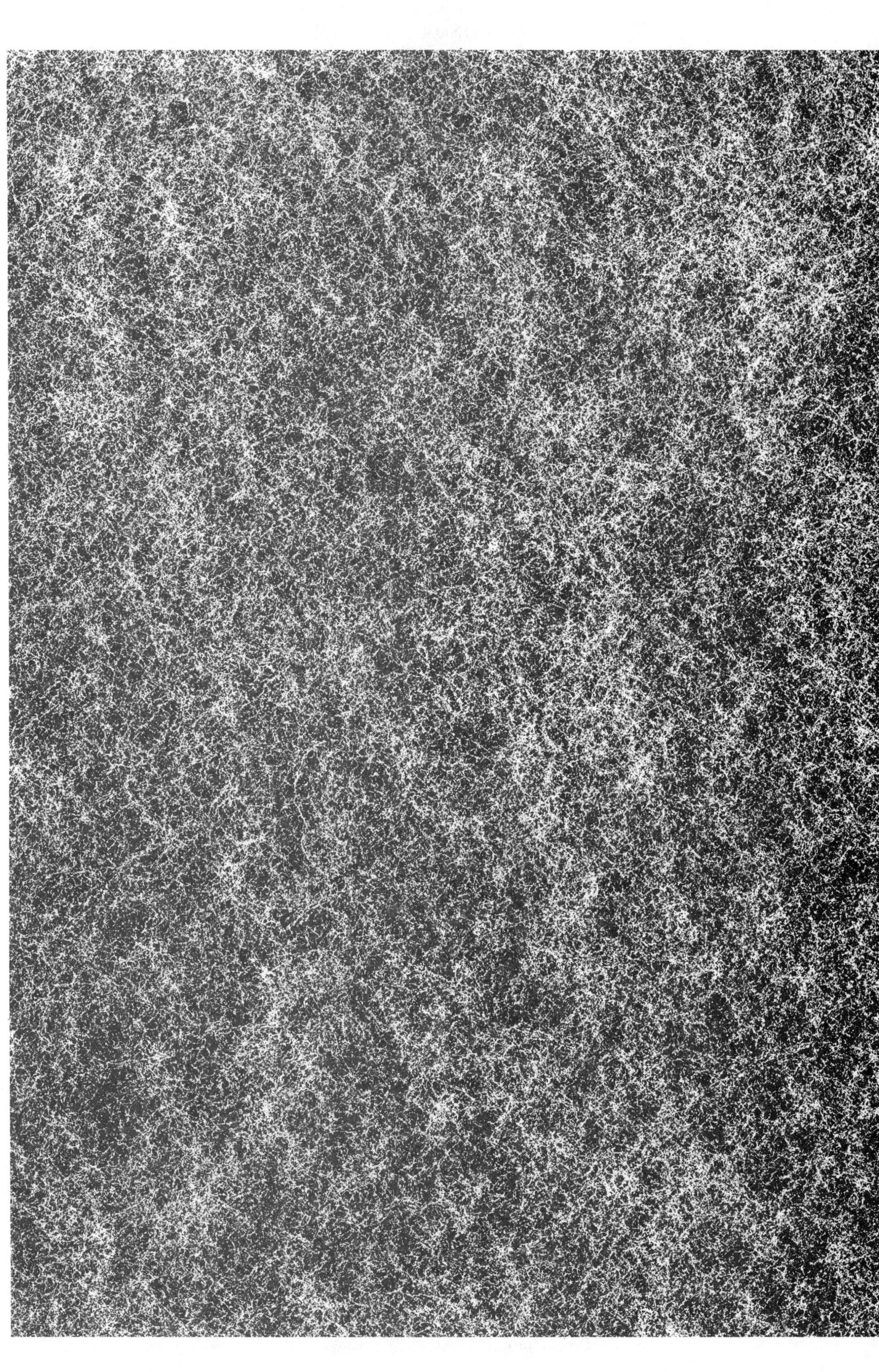

2 FLUID STATICS AND DYNAMICS

FLUIDS–1

The line between a low-pressure booster pump and a high-pressure boiler feed pump contains a running length of 65 ft of 6-in schedule-40 steel pipe, four long radius 90° elbows, a swing check valve, and a fully open gate valve. The feed pump is 12 ft higher than the booster pump in elevation. The flow rate is 740 gpm of 250°F water. The known pressures are

pressure before booster pump	37 psia
pressure before boiler feed pump	95 psia
pressure after boiler feed pump	1350 psia

(a) What is the pressure at the discharge of the booster pump?

(b) What is the brake horsepower requirement of the motor for the boiler feed pump if the pump's efficiency is 62%?

FLUIDS–2

A centrifugal pump has the following operating characteristics based on operation at 1800 rpm, 14.7 psia, and 85°F water.

Q (gpm)	H (ft)	NPSHR (ft)
550	50.0	8.0
600	47.5	9.5
650	45.0	11.1
700	42.1	13.0
750	39.1	15.0
800	36.0	17.2

(a) Tabulate the H and NPSHR characteristics as a function of Q if the pump is turned at 2000 rpm.

(b) Will the pump operate satisfactorily at 1800 rpm under the following conditions?

- 5000 ft altitude
- 90°F water
- 9-ft static discharge head
- 7-ft suction lift
- 650-gpm flow rate
- 10-ft friction loss in suction line

FLUIDS–3

In order to maintain the original appearances of a historic building, it has been decided to renovate the original hot-water radiator heating system. The building has a design heating load of 400,000 BTU/hr, which is satisfied by convective heat transfer from radiators throughout the building. Water leaves the boiler at 220°F saturated liquid and returns at 190°F. The hot water system has the following characteristics.

pipes
- $2\frac{1}{2}$-in steel, schedule-40
- 225 linear feet (exclusive of minor losses)

minor losses (all fittings are screwed)
- 8 90° regular elbows
- 1 swing check valve
- 2 gate valves (normally open)
- boiler heating coils head loss: 12 ft

pump

pump efficiency:	55%
motor efficiency:	75%

(a) What hydraulic horsepower is required?

(b) What is the motor's nameplate rating (in watts)?

FLUIDS–4

A pump was originally chosen based on an NPSHR of 21 ft for a given flow rate of 850 gpm of water. The water was to flow through 90 ft of 6.00-in (inside diameter) pipe (Darcy friction factor of 0.02) prior to entering the pump. Water entering the pump is currently at 180°F and 14.1 psia, and cavitation in the pump inlet is being experienced.

It has been proposed to enclose the inlet pipe with a larger diameter pipe for its entire 90 ft of run, and cold water circulated countercurrently through the outside pipe. The temperature at the pump inlet would be reduced to 145°F, with no change in flow rate or inlet pressure.

(a) What is the heat removal rate in BTU/hr?

(b) Determine analytically that the pump should cavitate under the original conditions.

(c) Determine analytically whether or not the pump will cavitate if the inlet is cooled.

FLUIDS–5

A large-diameter pipeline carrying natural gas must go under a river, and the pipeline engineer wants to anchor the pipeline to the bottom of the river bed. The pipeline will be completely buried in river-bottom mud as it crosses the river. The pipeline engineer has proposed placing a series of 12,000 lbm concrete weights at regular intervals along the pipeline to hold the pipeline down.

concrete	
density	140 lbm/ft^3
river	
depth (of burial)	22 ft
width	325 ft
river-bottom mud	
temperature	73°F
specific gravity	1.36 (saturated)
pipe	
material	steel
diameter	30 in
wall thickness	$\frac{1}{2}$ in
natural gas	
pressure	900 psia
molecular weight	18.9 lbm/pmole

Including an adequate safety factor, what should be the spacing between concrete weight locations along the submerged pipeline?

FLUIDS–6

During a fire, the fitting on a fire hydrant breaks and the fire hydrant flows freely from a 3.50-in (actual inside) diameter opening. The flow completely fills the outlet pipe.

coefficient of contraction	1.00
coefficient of velocity	0.82
coefficient of discharge	0.82

The centerline of the opening is 26 in from the ground. The average distance (i.e., centerline distance) between the fire hydrant opening and where the discharge stream meets the pavement is 8.5 ft.

(a) What is the flow (in gpm) out of the fire hydrant?

(b) What is the pressure (in psig) in the main?

FLUIDS–7

A 380-ft pipe (10 in, clean, flanged steel, schedule-40) carries 1900 gpm of 155 psig (at start of run), 90°F water. The drop in elevation between the entrance and exit is 45 ft. The pipe has the following features in its run.

10	long radius 90° els
4	standard radius 90° els
4	45° els
1	gate valve
1	globe valve
2	tees (through run)
1	tee (through stem)

(a) Disregarding entrance and exit losses, what is the head loss or gain (in feet of water)?

(b) Explain your answer to part (a).

FLUIDS–8

An installation of stainless steel pipe is being designed to carry a concentrated metallic salt solution. To prevent a spontaneous decomposition of the salt ions, the pressure in the solution must be kept at or above 18 psig.

pipe	
length	1500 ft
minor losses	2 wide open gate valves
	2 long radius 90° elbows
salt solution	
flow rate	250,000 lbm/hr
pressure at pipe inlet	23 psig
specific gravity	1.45
temperature	85°F
viscosity	1.20×10^{-4} ft^2/sec

(a) What is the minimum size of schedule-40 pipe that will satisfy the design requirement?

(b) What is the actual pressure drop (in psi) between the inlet and outlet of the pipe?

FLUIDS–9

The heat exchange section of a heat exchanger used to warm air is constructed as two multi-passage layers. 20 cfm of 5°C air enter the heat exchanger header. The loss coefficient for the supply header is 0.52. The loss coefficient for the receiving header is 0.9. All construction is smooth stainless steel. What is the friction loss (in inches of water) through the heat exchanger?

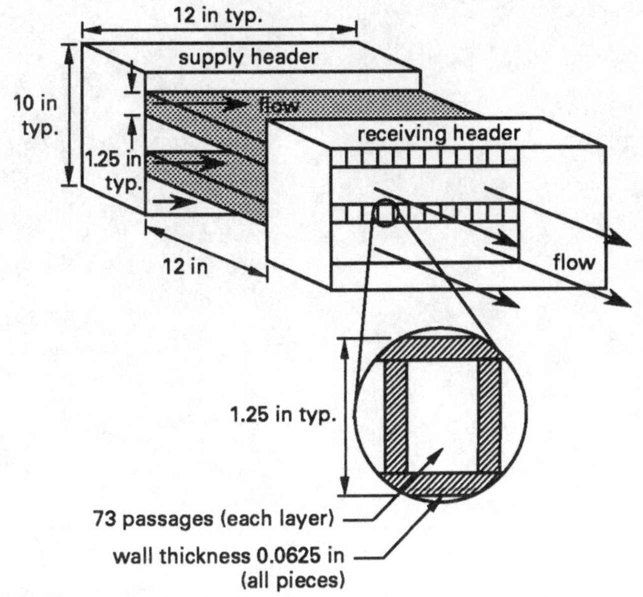

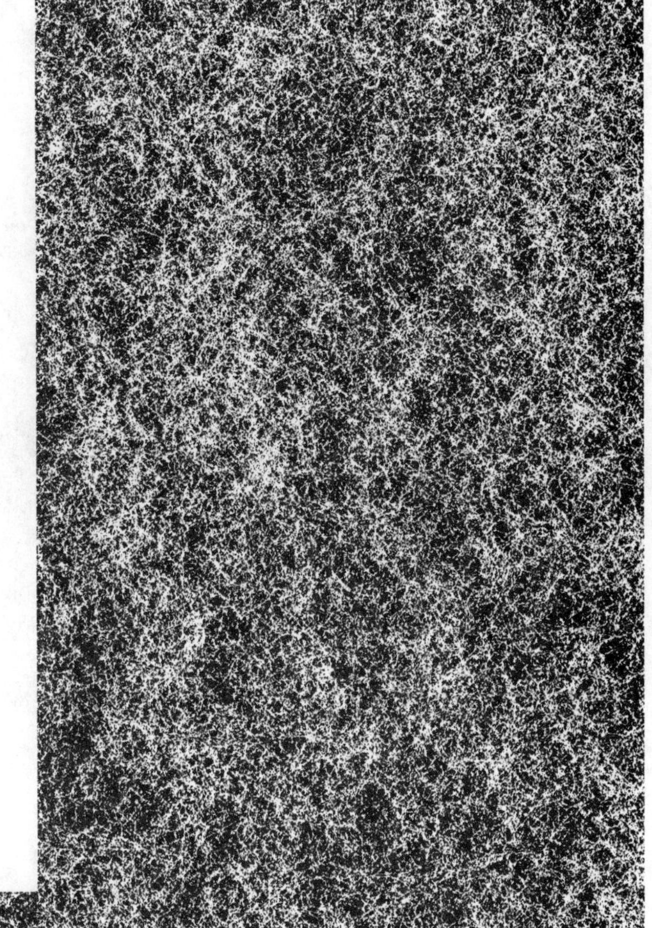

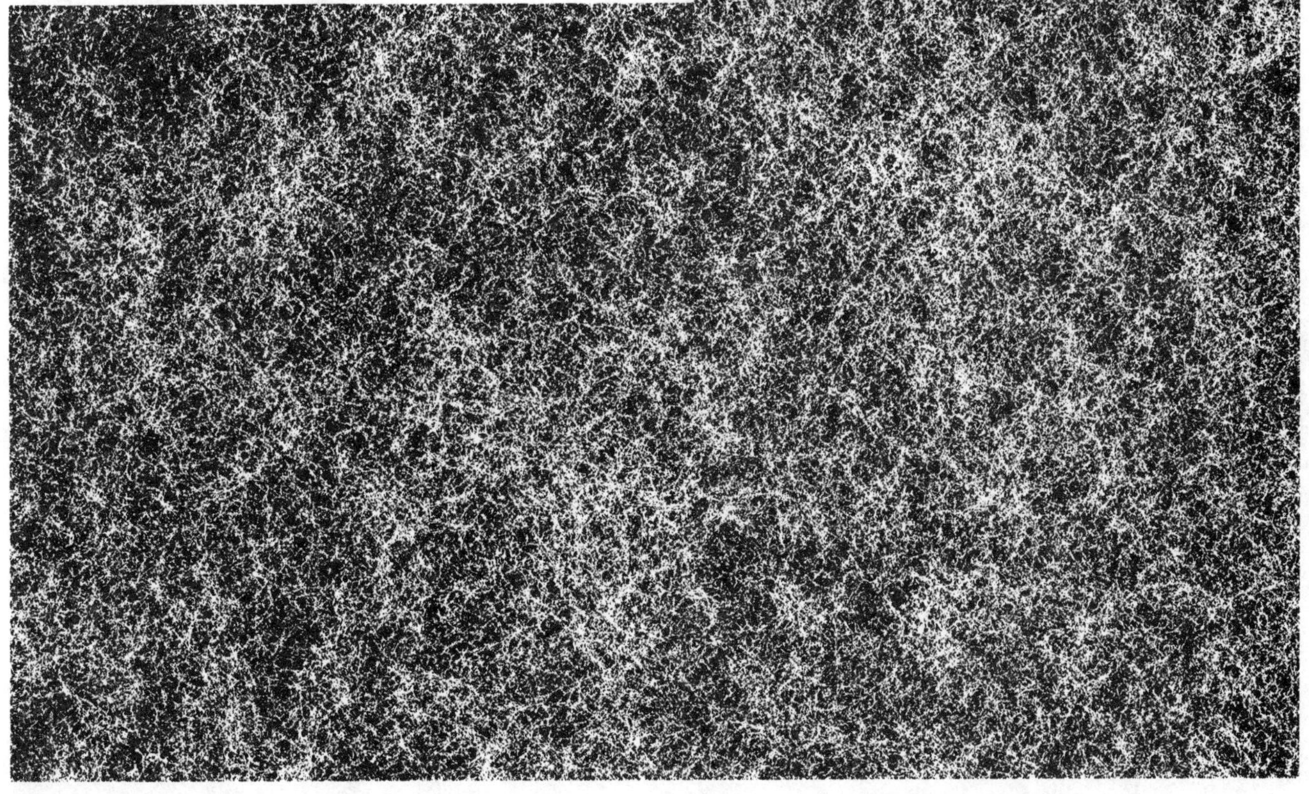

3 THERMODYNAMICS

THERMODYNAMICS–1

A low-humidity clean room receives some of its mechanical energy from a gas turbine. Gas flowing through the turbine is a mixture of carbon dioxide and nitrogen. The following data is available.

turbine mechanical efficiency	85%
turbine isentropic efficiency	70%
turbine power output	5.2 bhp
gas mixture flow rate	340 lbm/hr
turbine inlet gas temperature	240°F
turbine exit gas pressure	16 psia
gas mixture properties	
specific gas constant	0.057 BTU/lbm-°R
specific heat at constant pressure	0.218 BTU/lbm-°R
ratio of specific heats	1.355

(a) What is the turbine inlet pressure?

(b) What is the efficiency of the power generation process?

(c) What is the entropy change of the gas mixture?

(d) Sketch (and label) the T-s diagram for the expansion process.

THERMODYNAMICS–2

A waste heat boiler produces 700 psia (dry, saturated) steam from 220°F feedwater. The boiler receives energy from 40,000 lbm/hr of 1750°F (dry) air. After passing through the waste heat boiler, the temperature of the air has been reduced to 180°F above the steam temperature.

(a) How much steam is produced per hour?

(b) After expanding through a turbine, 80% of the condensate (saturated liquid at 190°F) is returned to a deaerator-heater. What quantity of boiler steam is required to heat the feedwater to 220°F? Boiler blowdown is negligible. 60°F make-up water is available to replace the deficit.

(c) How much make-up water (in lbm/hr) is required under the conditions of part (b)?

THERMODYNAMICS–3

A feedwater heater produces an unknown amount of 212°F (saturated, liquid) water. There is a heat loss of 225,000 BTU/hr from the exposed surfaces of the heater. The heater receives three input streams.

1. 4000 lbm/hr of saturated (liquid) water at 50 psia
2. 78,000 lbm/hr of 160°F saturated (liquid) condensate
3. an unknown quantity of 15 psia steam (quality of 80%)

(a) What is the flow rate (in lbm/hr) of the 15 psia steam?

(b) How much feedwater (in gpm) is produced?

(c) If the heat loss is reduced from 225,000 BTU/hr to 110,000 BTU/hr, how much feedwater (in gpm) will be produced?

THERMODYNAMICS–4

A steam turbine is powered by a set of fixed orifice nozzles. Each nozzle has an isentropic efficiency of 90%. 350°F (dry, saturated) steam at 280 ft/sec enters the nozzles. The steam expands adiabatically to 1750 ft/sec.

What is the (a) enthalpy, (b) quality, and (c) pressure of the steam as it leaves the nozzles?

THERMODYNAMICS–5

An unoccupied, pressure-tight equipment room for a nuclear power plant has a volume of 30,000 ft^3. The room is maintained at 14.7 psia and 80°F. In order to protect electronic equipment in the room from fire, a Halon 1301 extinguishing system has been installed. In the event of a fire, 2300 ft^3 of Halon (evaluated at 80°F and 14.7 psia) are injected into the room. None of the original air is displaced. Halon has a molecular weight of 150 lbm/pmole. Ideal gas relationships and perfect mixing at room temperature can be assumed.

(a) What is the pressure in the room after the Halon is injected?

(b) If the building is designed for a 1-in water gage allowable overpressurization (based on 14.7 psia), will the building be able to withstand the added Halon pressurization? Explain your answer.

(c) How would you vent the room if overpressurization is unacceptable?

THERMODYNAMICS–6

A steam turbine drives an electrical generator. Some of the steam is extracted before complete expansion to the condenser pressure, and the bleed steam is used for process heating. The remaining steam enters a condenser.

entering steam conditions	
flow rate	220,000 lbm/hr
pressure	1100 psia
temperature	900°F
bleed (extraction) steam conditions	
flow rate	110,000 lbm/hr
pressure	200 psia
condenser conditions	
pressure	2 in Hg (absolute)
equipment efficiencies	
turbine, isentropic	75%
generator, electrical	98%

(a) What is the temperature of the extracted steam?

(b) What is the enthalpy of the extracted steam?

(c) What power (in kW) is produced by the generator?

THERMODYNAMICS–7

A company has an unlimited source of 220°F, 25 psia water. Currently, the water is being discarded. The company also has an unfilled need for 25,000 lbm/hr of 300°F, 68 psia water. It has been proposed to produce the needed 300°F water in the following manner.

(1) The 220°F water will be throttled to 10 psia. During the throttling process, it is predicted that the water will partially vaporize.

(2) The water and vapor will enter a flash vessel (maintained at 10 psia) where the liquid and vapor will be allowed to separate by gravity.

(3) The 10 psia (saturated) liquid water will be drawn off and discarded.

(4) The 10 psia (saturated) vapor will be compressed to 70 psia by an 85% efficient compressor. The compressor will be powered by an electrical motor.

(5) The compressed vapor will pass through a heat exchanger where the temperature will be reduced to 300°F, and the pressure will be reduced to 68 psia.

(a) What will be the compressor's power requirement (in kW)?

(b) Is this a credible, valid process?

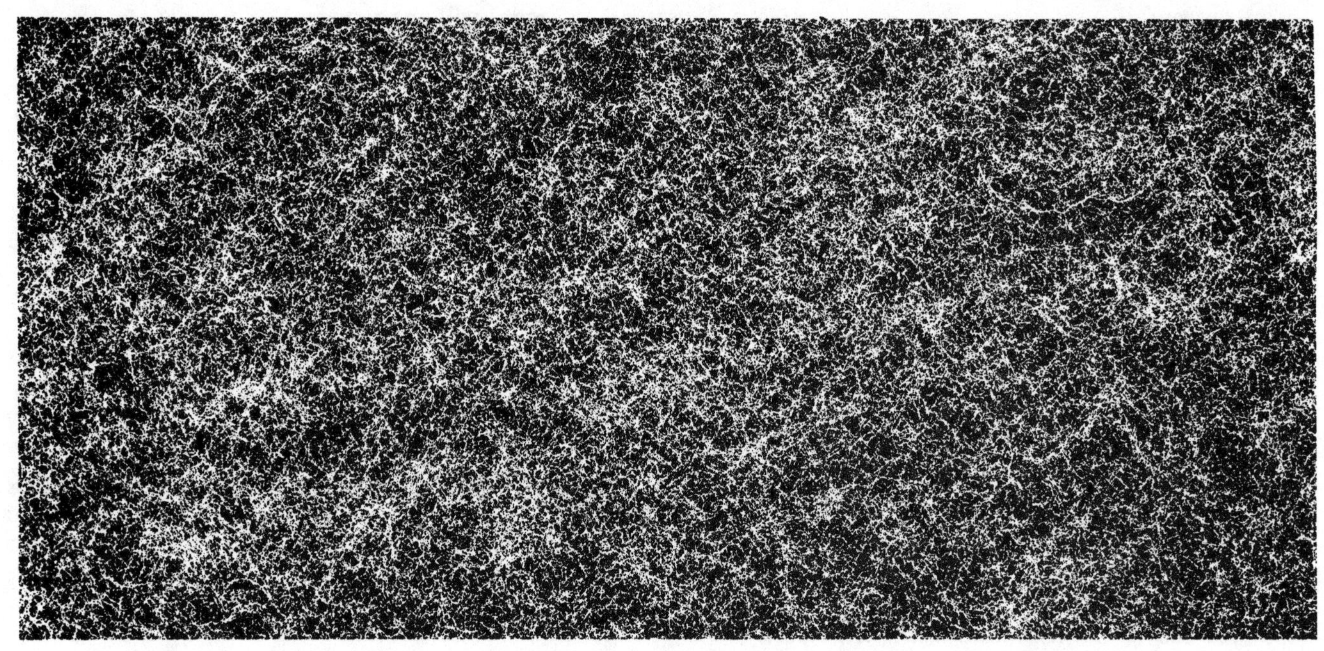

4 POWER CYCLES

POWER CYCLES–1

A company generates 1000 brake horsepower from a gas turbine. 9 lbm/sec of compressed combustion gases (specific heat at constant pressure of 0.239 BTU/lbm-°F) at 1640°F enter the turbine and exit at 1040°F. For years, the company has thrown away the gases leaving the turbine. Now, it is proposed to use the waste heat to generate steam for a low-pressure steam turbine.

A 90% efficient counterflow waste-heat boiler drops the combustion gases to 650°F in the process of producing 1000°F steam at 900 psia. An 80% efficient turbine expands the steam to 2 psia (the condenser condition). A 60% efficient boiler feed pump forces the condensate back into the waste heat boiler.

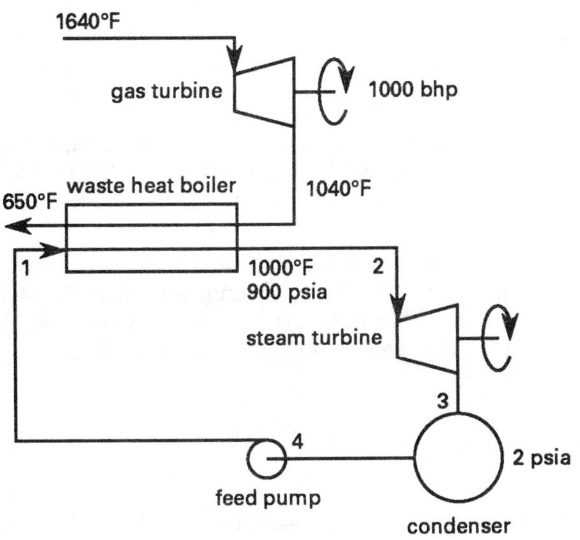

(a) What is the gross increase in power output (in horsepower)?

(b) What is the steam flow rate (in lbm/sec)?

POWER CYCLES–2

Currently, the expanded steam from a high/low pressure turbine set passes through a condenser cooled by river water. Environmental concerns have made it necessary to replace the river water with chilled water from a cooling tower. However, the condenser will be derated to operate at 2 psia instead of the current 1 psia. Saturated liquid leaves the feedwater heater.

Neglecting the pressure drop in the pipes, what will be the combined (both turbines) percentage decrease in output power? (All mechanical efficiencies are 100%. Disregard the insignificant work of the condenser evacuating pump.)

location	temperature, °F	pressure, psia
1	900	1000
2		40
3	900	
4		2
5		2
6		30
7		30
8		1000
9		70

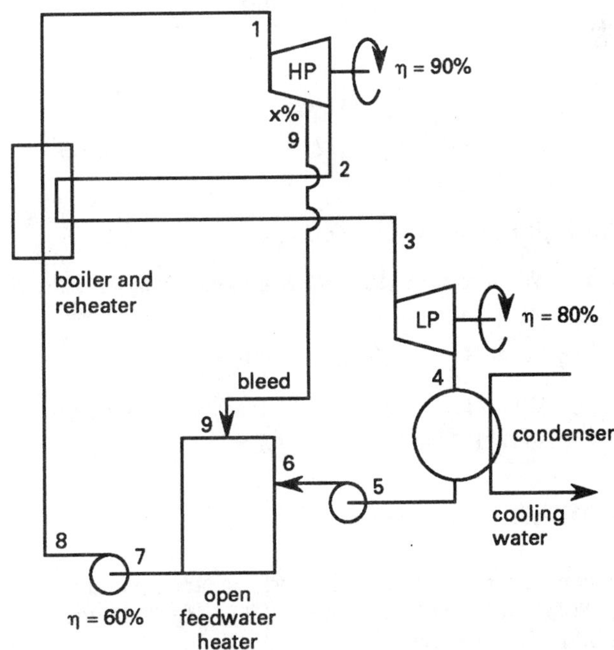

POWER CYCLES–3

A 12 MW (nominal) cogeneration (combined cycle) plant uses waste energy from a gas power turbine to drive a steam turbine. (Refer to the figure.) The following information has been compiled.

location	temperature, °R	pressure, psia
1	500	14.7
2		
3	2400	
4		18
5	760	
6	1460	300
7		10
8		10

net gas turbine generator output	5000 kW
fuel mass flow	8500 lbm/hr
fuel lower heating value	18,000 BTU/lbm
compression ratio	10:1
compression efficiency	85%
gas turbine efficiency	65%
generator efficiency	97%
steam turbine efficiency	80%
feed pump efficiency	60%
heat exchanger efficiency	85%

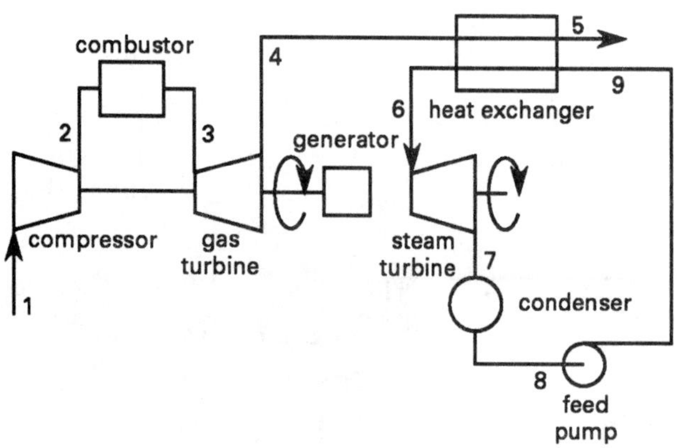

State all necessary assumptions.

(a) What power (in kW) is generated by the steam turbine?

(b) What is the combustor efficiency?

(c) What is the overall efficiency of this combined cycle process?

POWER CYCLES–4

A condenser in a typical Rankine steam cycle operates at 80°F and with a vacuum of 27 in Hg (referenced to a 30-in Hg barometer).

(a) Sketch the T-s diagram of such a typical Rankine cycle, showing the detrimental effect of a large amount of noncondensables.

(b) What is the percentage (by weight) of noncondensables in the condenser?

POWER CYCLES–5

It has been determined experimentally that the power needed by a particular car can be estimated by the following equation.

$$bhp = (4.3 \times 10^{-3}) \times S \times [0.016W + FA(S^2) \pm GW]$$

W = weight of car in pounds
F = streamline factor
A = projected frontal area in square feet
S = speed in mph
G = decimal grade (slope)

A particular car (standard transmission) has been able to attain a speed of 65 mph while coasting downhill in neutral, under the following measured conditions.

W = 2100 pounds
A = 20 ft^2
G = 7% downhill

(a) What size engine (bhp) should be installed to allow this car to maintain a speed of 55 mph while traveling up a 6% grade? (Include 20% extra for air conditioner, lights, and other accessories.)

(b) Gasoline has a density of 5.8 lbm/gal. If the brake specific fuel consumption (BSFC) under the conditions in part (a) is 0.43 lbm/bhp-hr, what mileage (in mpg) will this car achieve?

POWER CYCLES–6

A power generating plant satisfies a 15 MW (nominal) electrical demand by expanding steam through a turbine (mechanical efficiency of 98%, isentropic efficiency of 80%). The generator's electrical conversion efficiency is 99%. The plant also needs 55,000 lbm/hr of process steam for feedwater heating. There are two options for obtaining the steam from an existing supply line. (Refer to the illustrations for the operating conditions.)

option 1: Split the supply line entering the turbine.

option 2: Bleed steam out of the turbine.

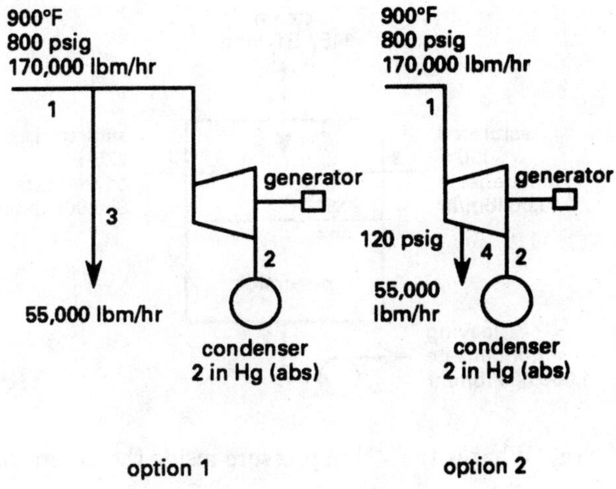

(a) What are the steam enthalpies at the condenser inlets for both options?

(b) What is the difference in power output between the two options? Express your answer in MW.

POWER CYCLES–7

A cascade refrigeration machine uses refrigerant R-12 in the upper cascade and refrigerant R-22 in the lower cascade. 85°F water is available for heat removal in the condenser of the upper cascade. The cooling effect is obtained at −50°F.

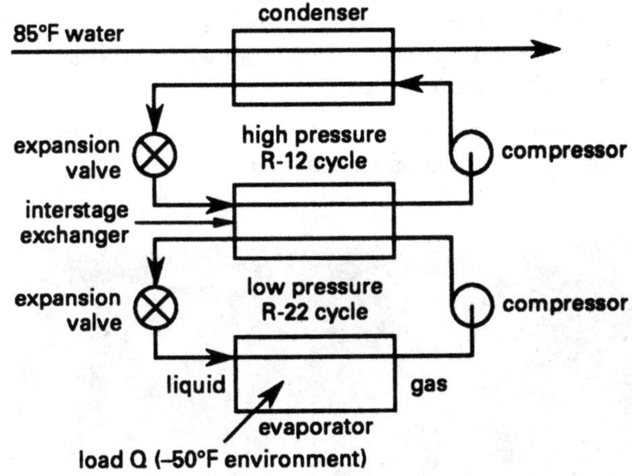

(a) Estimate reasonable temperature differences in the three heat exchangers: condenser, interstage, and evaporator.

(b) Using actual property data for the two refrigerants, calculate the pressures at the entrances to the three heat exchangers: condenser, interstage, and evaporator.

POWER CYCLES–8

A cogeneration plant operates on the cycle shown. For the purpose of this problem, the gas can be considered to be air.

compressor isentropic efficiency	85%
gas turbine expansion ratio	10:1
gas turbine isentropic efficiency	74%
steam turbine isentropic efficiency	82%
pump work	negligible

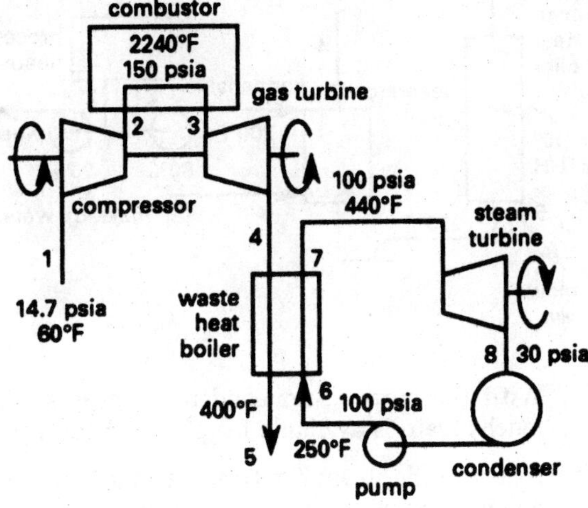

(a) What is the overall efficiency of this cycle?

(b) What is the ratio of steam mass flow rate to gas mass flow rate?

POWER CYCLES–9

An R-12 refrigerator incorporates a reciprocating compressor. The following information is known.

refrigerant vapor	
k (ratio of specific heats)	1.3
compressor	
volumetric clearance	6%
rotational speed	125 rpm
overall thermodynamic efficiency	60%
polytropic exponent	1.3
cycle	
refrigeration effect	8 tons
condenser pressure	120 psia
evaporator temperature	40°F
environment	
temperature	80°F
pressure	14.7 psia

(a) What is the piston displacement?

(b) What is the input power (in kW)?

POWER CYCLES–10

Currently, a waste-heat boiler uses 2.4×10^8 BTU/hr of waste heat to produce 180 psia (dry, saturated) steam for process heating. Currently, no condensate is returned to the boiler. It has been proposed to add a deaerator, and to return 80% of the steam to the deaerator. The remaining 20% of the flow will be replaced with 50°F make-up water. (See process configuration illustration.)

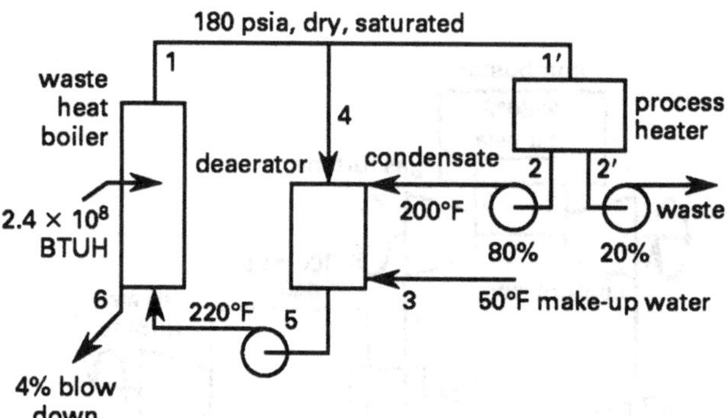

(a) With the current, noncondensing process, how much steam is available for process heating?

(b) If 80% of the steam is returned to the boiler, how much steam will be available for process heating?

POWER CYCLES–11

A deaerator operates with a 0°F hot well depression. (That is, the temperature of the condensate removed is equal to the steam inlet temperature.) The deaerator is part of a power plant that produces 250 MW of electricity when operating at 90% of its capacity in steady state and equilibrium conditions. The condensate pump supplies negligible energy.

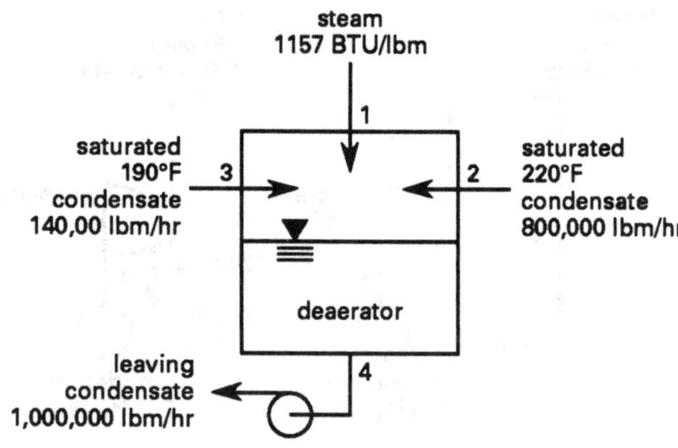

(a) What is the steam pressure inside the deaerator?

(b) What is the quality of the entering steam?

POWER CYCLES–12

A jet engine is modeled as a serial combination of a compressor; an adiabatic, constant pressure combustor; and a turbine (in that order). Air (considered to have a variable specific heat) is compressed through a compression pressure ratio of 8.25:1. Air enters the compressor at 14.7 psia and 60°F. Air leaves the combustor at 2200°R. The compression isentropic efficiency is 0.83, and the turbine isentropic efficiency is 0.85.

An engineer claims that 5% of the air should be bypassed around the compressor and combustor and used to cool the engine walls in a double-wall arrangement. If that is done, the air leaving the combustor can be increased in temperature 250°F, and the combustor pressure increased by 5 psi. It is claimed that the engine will produce more thrust.

Assuming that the bypassed air produces no thrust, and ignoring fuel mass, is the engineer correct?

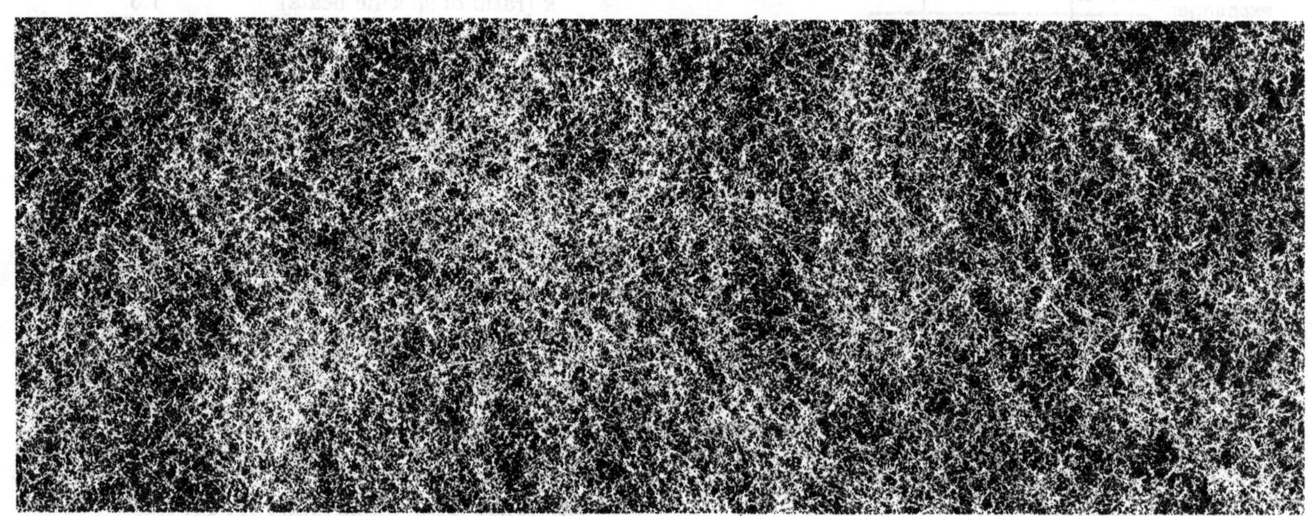

5 COMPRESSIBLE FLUID FLOW

COMPRESSIBLE FLUID FLOW–1

Air ($k = 1.40$) flows in a converging/diverging nozzle constructed with a design pressure ratio (i.e., ratio of exit pressure to stagnation pressure) of 0.1278. There is a uniform divergence over the entire length, L, of the nozzle diverging section.

(a) At what minimum exit pressure ratio (in terms of the chamber stagnation pressure) does a shock wave stand at the exit of the nozzle?

(b) At what distance (in terms of the diverging section length, L) from the throat does the shock wave occur if the ratio of exit pressure to chamber stagnation pressure is 0.68?

COMPRESSIBLE FLUID FLOW–2

A sugar beet plant needs to install steam humidifiers at various locations to keep the sugar beets from drying out and shriveling up.

outdoor design conditions	
dry bulb temperature	15°F
relative humidity	10%
indoor design conditions	
dry bulb temperature	70°F
relative humidity	40%
pressure	14.7 psia
plant building	
volume	450,000 ft^3
occupancy	6 people seated
ventilation	
air changes per hour	3
make-up air from outside	25,000 cfm
moisture absorbed by sugar beets	350 lbm/hr
steam humidifiers	
steam pressure	25 psig (saturated)
humidifier orifice size	0.37 in

(a) How much steam (in lbm/hr) is required for humidification?

(b) How many humidifier orifices should be installed around the plant?

COMPRESSIBLE FLUID FLOW–3

A high-temperature, gas-cooled nuclear reactor (HTGR) vessel contains 900 ft^3 of hydrogen at 600°F and 1100 psia. The vessel is surrounded by a very large containment building filled with air at 14.7 psia and 200°F. The reactor vessel sustains damage (of an unknown nature), and the hydrogen begins leaking out through a short length of 1-in (inside diameter) pipe that has been crushed to an effective orifice size of $\frac{1}{2}$ in. Due to the size and controlled venting of the containment structure, the pressure remains at 14.7 psia during the entire hydrogen discharge. Emergency equipment replaces (displaces) the lost hydrogen with nitrogen, maintaining the original temperature and pressure in the reactor vessel. (No nitrogen is vented into the containment building.) Assume a discharge coefficient, and determine how long it will take to vent all of the hydrogen into the containment building.

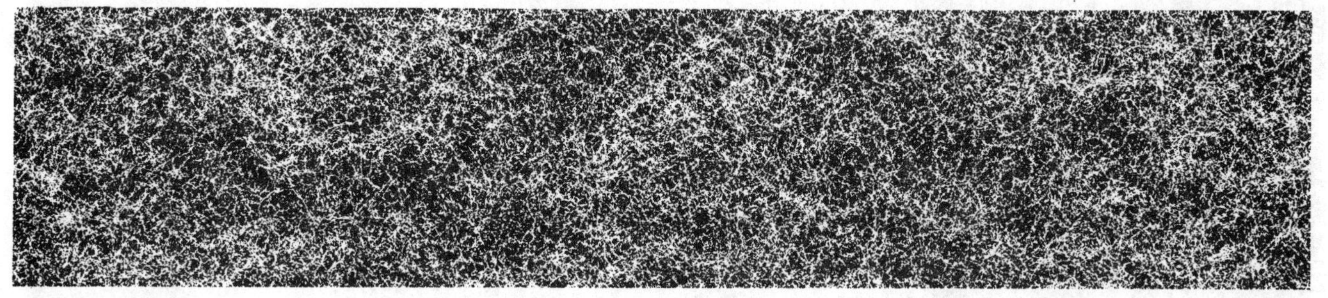

6 COMBUSTION

COMBUSTION–1

Propane (C_3H_8) has been chosen as the fuel for a new furnace. The fuel is combined with atmospheric air supplied at 14.7 psia, 80°F dry bulb, and 70°F wet bulb. Orsat tests have been taken, and the products of combustion have been determined as follows.

product	% by volume (dry basis)
CO_2	9.9
O_2	6.1
CO	0.0
N_2	84.0
SO_2	0.0

The furnace stack is very high, and it is feared that the combustion gases will cool significantly as they rise. It is desired to keep the furnace combustion gases at least 250°F hotter than the water vapor's dew point temperature. If the pressure in the stack is 14.5 psia, what is the minimum temperature to which the combustion gases can be cooled?

COMBUSTION–2

An internal combustion engine completely burns a mixture of (vaporized) hydrocarbon fuel and dry, desert air. Where sampled, the exhaust gas is at 15 psia and 120°F, and consists of the following products of combustion.

product	% by volume (dry basis)
CO_2	11.1
CO	5.3
CH_4	1.3
H_2	2.1
O_2	0.4
N_2	79.8

(a) What is the dew point of the water vapor in the exhaust?

(b) What is the actual air/fuel ratio (in lbm/lbm)?

COMBUSTION–3

A diesel-type engine burning dodecane ($C_{12}H_{26}$) is installed in a military combat vehicle expected to operate under smokey and dusty (but dry) desert conditions. At full load, the engine produces 625 bhp while consuming 0.43 lbm of fuel per bhp per hour. The throttle has been designed to admit excess air according to the following power curve.

$$\% \text{ excess air} = 700\% - 6 \times (\% \text{ of full load throttle})$$

The subcontractor supplying air filters for the combat vehicle has recommended that the air velocity through the filter be limited to 600 ft/min. Desert air is 14.7 psia and 90°F.

If the engine is fully loaded, what filter area is required?

COMBUSTION–4

A mixture of gases is burned with dry air. The burner inlet operates with a negative gage pressure, and the fuel/air mixture is at 60°F and 12 psia. The fuel consists of the following compounds.

fuel compound	molar fraction (moles/mole mixture)
CH_4	0.87
C_2H_6	0.04
C_3H_8	0.02
C_4H_{10}	0.005
C_5H_{12}	0.002
N_2	0.063

(a) What is the stoichiometric air/fuel ratio (in lbm/lbm)?

(b) Assuming stoichiometric (complete) combustion and dry air, what is the dew point of the stack gases if the stack pressure is 14.4 psia?

(c) What is your recommendation for the stack gas temperature? (Explain your answer.)

COMBUSTION–5

Air at 60°F and 14.7 psia is combined with 40 lbm/min of fuel oil ($C_{16}H_{32}$, density of 8.23 lbm/gal, and heating value of 152,000 BTU/gal) to provide combustion energy for terra cotta pot glazing. The furnace temperature (maintained by combustion gases at the same temperature) has dropped to 1350°F due to a large intake system leak which is admitting excess air. The average molecular weight of the combustion gas is 32.4 lbm/pmole, and the average ratio of specific heats (k) is 1.40.

(a) What is the stoichiometric air (in ft^3/min) required, excluding excess and leakage?

(b) What mass (in lbm/lbm of fuel) of combustion gases is produced per minute?

COMBUSTION–6

The combustor of a gas turbine receives 11.4 lbm/sec of 575°F, 145 psia (stagnation properties) air. Liquid isooctane at 74°F is added, and combustion is complete. The temperature of the combustion gases leaving the combustor is 2250°F. State your assumptions, and find the air/fuel ratio in lbm/lbm.

COMBUSTION–7

A producer of gourmet popcorn has leftover corncobs coming out its ears. The producer has decided to try to heat part of the packaging plant during the winter with sun-dried corncobs leftover from the summer (and stored outside). The following information is available.

outside design air conditions	
pressure	14.7 psia
temperature	34°F
moisture content	negligible

corncob ultimate analysis	
C	48%
H	5%
O	40%
N	1%
ash	6%

corncob heating value (as fired) 6600 BTU/lbm

combustion losses	
unburned solid fuel in ash	negligible
ash heat retention	negligible
conduction, radiation losses	negligible

(a) Neglecting the heating effect (to the plant) and any minor losses, what percent excess air is required to limit the exhaust temperature to 450°F?

(b) Using your answer from part (a), what volume (in ft^3) of air is needed to burn 500 lbm of corncobs?

7 HEAT TRANSFER

HEAT TRANSFER–1

The State Agricultural Commissioner has ordered that apples be quarantined and kept in cold storage for 3 months to kill suspected Mediterranean fruit fly larvae. The apples are brought in from the field at 80°F, and are placed in single layer trays with the fruit spaced far enough so that no apple affects another. The trays ride a slow conveyor which brings the apples into contact with 14.7 psia, 10°F air. Cooling is by natural convection only. When the apple cores reach 40°F, the apples are transferred to a cold fruit locker (maintained at 40°F). The apples can be assumed to be homogeneous 3.5-in diameter spheres with the following properties.

conductivity (k)	0.42 BTU-ft/hr-ft^2-°F
film coefficient (h)	5.7 BTU/hr-ft^2-°F
specific heat (c_p)	1.0 BTU/lbm-°F
thermal diffusivity (a)	0.0063 ft^2/hr

How long will it take for the apple centers to reach 40°F?

HEAT TRANSFER–2

An experimental, power-rectifying diode dissipates 5 watts of electrical energy from its casing, and is cooled by a combination of natural convection and radiation (emissivity of 0.65). For initial heat transfer predictions, it has been decided to model the diode as an upright circular cylinder, 0.8 in in diameter and 1.7 in high. The base of the diode is secured in a glob of cured silicone, which effectively insulates that surface. For the purpose of both convection and radiation, the environment within the electronics enclosure can be considered to be at 120°F and 14.7 psia.

(a) What is your prediction of the diode's surface temperature?

(b) What percentage of the heat loss is by convection?

HEAT TRANSFER–3

400,000 lbm/hr of 270°F water are to be heated to 370°F by condensing saturated 390°F steam. (The steam does not subcool upon condensing.) Your job is to design the heat exchanger. The following parameters represent constraints on your design.

number of tube passes	4
tube fluid	water being heated
number of shell passes	1
shell fluid	steam being cooled
U (based on outside tube area)	700 BTU/hr-ft^2-°F
tube outside diameter	1 in
tube wall thickness	$\frac{1}{16}$ in
bulk tube fluid velocity	5 ft/sec

(a) How many tubes are required per pass?

(b) How many tubes are required total?

(c) Allowing 3 ft for headers and flanges, how long should the exchanger be?

HEAT TRANSFER–4

A poor heat-conducting copper alloy (k = 45 BTU-ft/hr-ft^2-°F) is drawn into 0.012-ft diameter wire for a critical aerospace application in an electronics enclosure. The conditions within the enclosure are maintained at 90°F by a forced air cooling system. To keep the copper alloy below 270°F and to maximize heat transfer, the wire is to be covered with insulation (k = 0.03 BTU-ft/hr-ft^2-°F). The effective film coefficient for the insulation has been conservatively estimated as 2.0 BTU/hr-ft^2-°F. What insulation thickness should be used over the wire to maximize heat transfer?

HEAT TRANSFER–5

Inmates of a minimum-security prison are complaining of being too cold, but the warden is reluctant to install individual heaters in each cell. It has been proposed to replace a 2-ft wide × 5-ft high section of wall between cells with $\frac{1}{4}$-in mild steel plate. Low voltage current will flow through the steel panel, generating 200 watts of resistance heating. The steel plate will be covered on both sides with 1-in sheetrock (gyp-board), providing two heating surfaces to rooms which are to be maintained at 70°F. For the sheetrock,

conductivity (k)	1.0 BTU-in/hr-ft^2-°F
emissivity (ϵ)	0.90

(a) What is the temperature at the interface between the steel plate and the sheetrock?

(b) What is the maximum steel temperature?

(c) What is the temperature at the outside surface of the sheetrock?

HEAT TRANSFER–6

Multi-Bend Company always tests its heat exchangers prior to shipping them to customers. In one case, a heat exchanger was tested as follows.

tube passes	2
tube area	60 ft^2
tube fluid	water being heated
tube fluid flow rate	120 gpm
tube fluid temperature in	70°F
tube fluid temperature out	150°F
shell passes	1
shell fluid	steam being cooled
shell fluid temperature in	240°F (saturated vapor)
shell fluid temperature out	240°F (liquid and vapor)

After being placed in service for 2 years, the customer has reported a change in the heat exchanger's performance. The current performance appears to be as follows.

tube fluid	water being heated
tube fluid flow rate	100 gpm
tube fluid temperature in	70°F
tube fluid temperature out	130°F
shell fluid	steam being cooled
shell fluid temperature in	250°F (saturated vapor)
shell fluid temperature out	250°F (liquid and vapor)

What is the current value of the fouling resistance (in hr-ft^2-°F/BTU)?

HEAT TRANSFER–7

A utility company has a policy of insulating all large-diameter pipes carrying steam. Insulation, where used, has been standardized as 3 in of 85% magnesia covered by a 1-in thick powdered diatomaceous earth blanket. Utility company buildings are normally maintained at 70°F. The utility operates 24 hours per day, every day of the year.

(a) What is the percentage savings in heat loss obtained by insulating a 10-in diameter, schedule-160 steel pipe (oriented horizontally) carrying 400°F turbulent steam?

(b) The utility company currently spends $2.46 per million BTU generated. Its effective annual interest rate is 20%. What is the maximum amount of money it should spend on insulating the pipe in part (a)?

HEAT TRANSFER–8

To prevent electromagnetic interference, a 200-watt electronic device has been enclosed with a 45-cm × 30-cm (footprint) × 25-cm high aluminum box placed over it. (No heat transfer occurs through the base.) The aluminum box is electrically grounded. Aluminum has an emissivity of 0.7. The box reaches a steady uniform (unknown) temperature. The local environment is at 20°C.

(a) Neglecting radiation, what is the heat transfer (in watts) from each surface of the box?

(b) Neglecting radiation, what is the box surface temperature (in °C)?

(c) Including radiation and assuming that the average convective film coefficient for the top and sides is 4.883 kcal/hr-m^2-°C, what is the box surface temperature (in °C)?

HEAT TRANSFER–9

Water temperature is increased in a heat exchanger at the expense of a decrease in temperature of a non-foaming oil. The following information is known.

number of tube passes	2
tube (inside) area	40 ft^2
overall heat transfer coefficient, based on inside area	280 BTU/hr-ft^2-°F
tube fluid	water being heated
tube fluid temperature in	60°F
tube fluid flow rate	12,000 lbm/hr
number of shell passes	1
shell fluid	oil being cooled
shell fluid temperature in	280°F
shell fluid specific heat	0.60 BTU/lbm-°F
shell fluid flow rate	20,000 lbm/hr

What are the temperatures of the two fluids as they leave the heat exchanger?

HEAT TRANSFER–10

A catalyst in aqueous solution used in a polymerization process will begin to precipitate out if the solution temperature drops below 50°F. Unfortunately, fumes from this catalyst solution are toxic, and storage in an outside tank is unavoidable. The tank is constructed as a cube, 12 ft on each side. The tank walls consist of 14 in of high-strength concrete covered with 1-in steel plate for protection from impact. (The concrete and steel plate are in intimate contact.) The tank has been mounted over a thick elastomeric pad in order to eliminate heat transfer to the ground. On a worst-case day, the exterior air temperature will be −10°F. The average inside convective film coefficient is 500 BTU/hr-ft^2-°F.

(a) How much heat (in BTU/hr) must be added to the catalyst solution to maintain the solution at 50°F on a worst-case day with no wind?

(b) Describe how your analysis would differ (from what you did in part (a)) if the catalyst is to be kept above its freezing temperature of 28°F. (No calculations are required.)

(c) Describe how your analysis would differ (from what you did in part (a)) if there is a 25-mph wind. (No calculations are required.)

(d) Low pressure (5 psig) waste steam is available to heat the solution. However, there are no heating coils yet in the tank. How many feet of 1-in schedule-40 steel pipe should be installed?

HEAT TRANSFER–11

An intrusion-detection system partially relies on a 25-watt infrared beam emanating from a source in a 60°F room (at night) and aimed at a diffusing target (reflector) elsewhere in the room. The source box is a zinc (pot-metal) casting, but is to be modeled as a 5-in cube mounted to the room wall. There is no heat transfer to the wall by conduction, but radiation and convection (convective film coefficient of 1.5 BTU/hr-ft^2-°F for all exposed surfaces) take place from the remaining five sides. (Disregard the effect of the beam diffuser/lens opening.) The source box has an emissivity of 0.92. To protect burglars against accidentally burning themselves, the source box must be kept below 150°F. Will the source box surface reach 150°F?

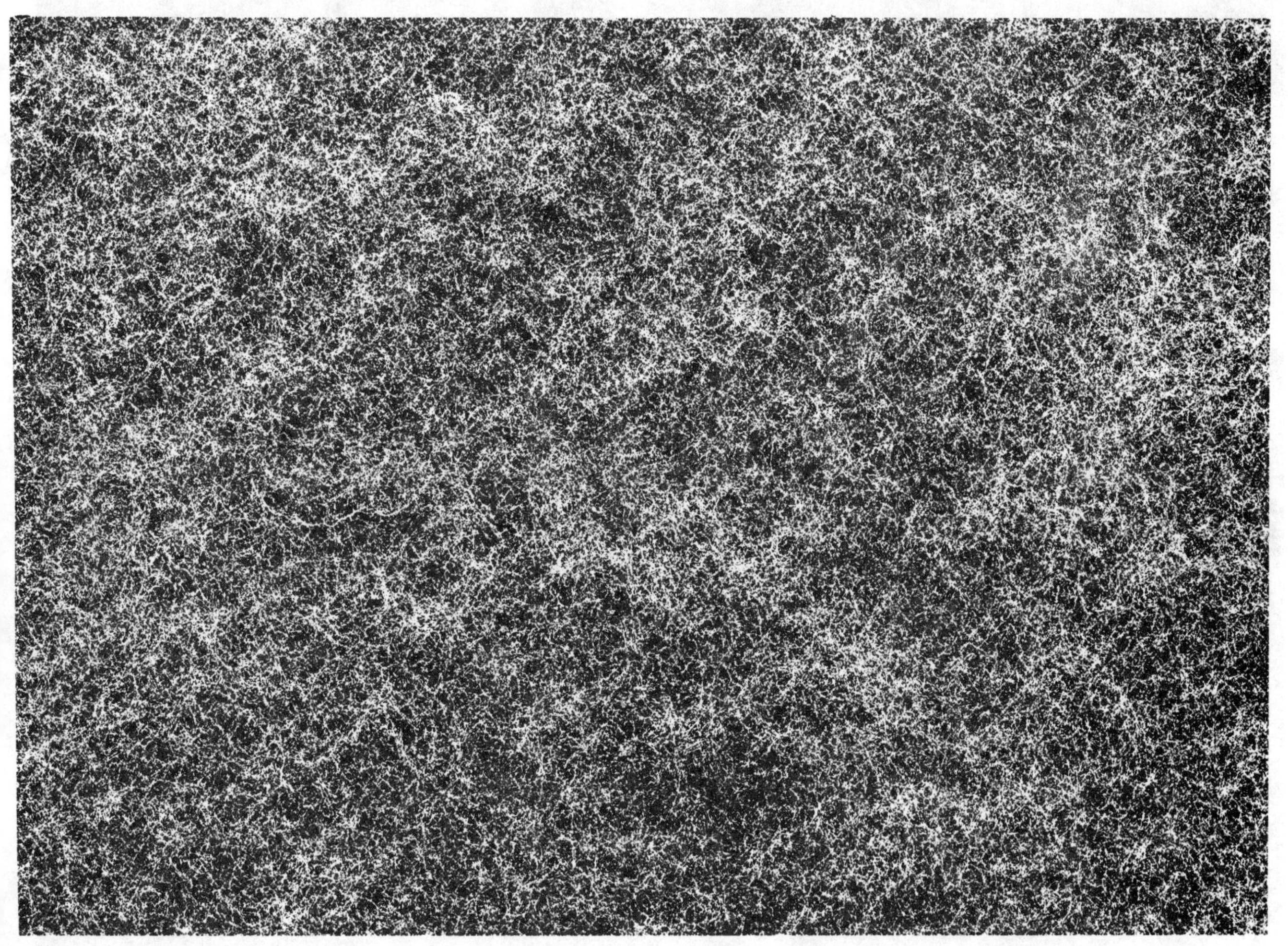

8 HEATING, VENTILATING, AND AIR CONDITIONING

HVAC–1

340 lbm/hr of dry 800°F air are diverted through a high temperature scrubber in order to reduce particulate emissions. Since the scrubber is constructed with elastromeric seals, the air temperature must first be reduced to 360°F. This temperature reduction occurs when the air is passed through a 78°F water spray. The pressure in the spray chamber is 85 psia.

(a) How much water (in lbm/hr) is required?

(b) What is the relative humidity of the air as it leaves the spray chamber?

HVAC–2

A building is located in a part of the country where the annual heating season is 23 weeks. The building is occupied from 8:00 a.m. until 6:00 p.m., Monday through Friday. The building owner pays $0.30 per therm for gas heating (which includes the costs of fans, pumps, etc.), and the gas furnace efficiency is 77%. The building has the following characteristics.

internal volume	600,000 ft^3
inside temperature	70°F
ventilation rate	
occupied	1 air change per hour
unoccupied	$\frac{1}{2}$ air change per hour
walls	
area	10,000 ft^2
U value	0.15 BTU/hr-ft^2-°F
windows	
area	2500 ft^2
U value	1.10 BTU/hr-ft^2-°F
roof	
area	25,000 ft^2
U value	0.06 BTU/hr-ft^2-°F
floor	
slab on grade	
exposed edge	720 lft
B value	1.6 BTU/hr-lft-°F

In the past, the building owner kept the interior temperature fixed at 70°F. What will be the annual savings (in dollars) if the building owner installs an automatic setback thermostat to reduce the interior temperature from 70°F to 58°F during unoccupied time?

HVAC–3

A building is located at 33°N latitude. At its location, the daily temperature swing is 24°F at 4:00 p.m. in mid-July (sun time). The inside design temperature is 78°F. The outside design temperature is 96°F. The building is constructed as follows.

walls
: 1700 ft^2 facing north
: 1500 ft^2 facing south
: 1600 ft^2 facing east (exclusive of windows)
: 1500 ft^2 facing west
: (from outside to inside)
  - 4-in brick facing
  - 4-in hollow clay tile
  - $3\frac{1}{2}$-in fiberglass insulation
  - 2 × 4 studs
  - $\frac{5}{8}$-in drywall gypsum

roof
: 6000 ft^2
: (from outside to inside)
  - 4-in concrete, natural color
  - 2-in elastomeric insulation ($k = 0.28$ BTU-in/hr-ft^2-°F)
  - asphalt-saturated felt paper, 4-ply
  - $\frac{1}{2}$-in plywood
  - $8\frac{1}{2}$-in fiberglass insulation
  - 2 × 10 ceiling joists
  - $\frac{5}{8}$-in drywall gypsum
  - acoustical ceiling tile, $\frac{1}{2}$ in

windows
: 150 ft^2 facing east (only)
: $\frac{1}{4}$-in thick, single glazing
: light-colored mini-blinds (full coverage inside)
: no exterior shading or overhangs

For the purpose of this problem, ignore heat transmission through studs, joists, and the floor. Ignore heat gain from lights and occupants.

(a) What is the instantaneous heat gain at 4:00 p.m. in mid-July?

(b) Why or why not is this the peak cooling load?

HVAC-4

An opera house with seating for 600 people (includes the orchestra and performers) is being retrofitted with air-conditioning according to the following parameters.

heat gain from outside	100,000 BTU/hr
additional heat sources	insignificant
indoor design conditions	
dry bulb temperature	75°F
relative humidity	50%
outdoor design conditions	
dry bulb temperature	90°F
wet bulb temperature	70°F
outside air per person	5 ft^3/min
dry bulb temperature of air entering the room	60°F
heat sink	
chilled water	
water temperature in	42°F
water temperature out	52°F
reheat	not permitted

State your assumptions.

(a) What air flow (in ft^3/min) should enter the room?

(b) How much chilled water (in lbm/hr) is required?

(c) What are the temperatures of the air entering and leaving the coil?

(d) What is the capacity of the air-conditioner in tons?

HVAC-5

A building uses an electrical heat pump for winter heating. However, for part of the winter, the capacity of the heat pump is insufficient, and pure electrical heating must be used to make up the difference. (No heating is required at 70°F or above.) The temperatures in the table are accurate to within 2.5°F.

design heat loss	3×10^5 BTU/hr
indoor design temperature	70°F
outside design temperature	0°F
internal heat sources	7500 BTU/hr
cost of electricity	$0.08/kW-hr

temperature (°F)	hours per year	heat pump capacity (MBH)	power to operate heat pump (kW)
65	820	60	6.0
60	600	55	5.8
55	450	50	5.6
50	340	45	5.4
45	250	40	5.2
40	190	35	5.0
35	130	30	4.8
30	90	25	4.6
25	60	20	4.4
20	30	15	4.2
15	15	10	4.0
10	3	5	3.8

(a) What is the total annual cost of the pure electrical heating? (Hint: MBH = thousands of BTU per hour.)

(b) What is the cost of running the heat pump?

HVAC-6

At a particular location, there are 220 days in the heating season, and the number of degree days is 4839. The design heating load for a building was originally calculated as 135,000 BTU/hr based on a 70°F interior design temperature and a 0°F exterior design temperature.

Over the years, the building occupants have tampered with the heating controls, and the building owner has determined that the interior temperature has been maintained at 73°F for several years.

(a) What is the current heat loss over the heating season?

(b) What savings (in BTU/yr) can the building owner expect if the interior temperature is reduced to 68°F?

HVAC-7

A 24-hour disco has a maximum occupancy of 240 adults (120 males, 120 females), half of which are usually dancing and half of which are talking while seated around tables. The disco experiences the following heat gains in addition to occupant loads.

food preparation moisture	45,000 grains/hr
heat gain and equipment heating	140,000 BTU/hr

60°F (dry bulb) air is supplied by the air-conditioning apparatus, and air is removed from the disco when it reaches 77°F (dry bulb) and 50% relative humidity. Make-up air consists of 4500 ft^3/min of outside air (90°F dry bulb, 76°F wet bulb).

(a) What volume (in ft^3/min) of air is supplied to the disco?

(b) What is the humidity ratio of the air supplied to the disco?

(c) What is the tonnage of the air-conditioning unit?

HVAC–8

A retail apparel goods store is maintained at 75°F and 40% relative humidity, although the exterior weather conditions are 25°F and 20% relative humidity. A 10-mph wind causes the building to lose 800 ft^3/min of inside air (which is replaced by outside air). There are typically 35 sales clerks and customers in the store at any given time. In addition to the occupant latent load, a humidifier supplies moisture to the air.

(a) How much moisture (in lbm/hr) should the humidifier supply?

(b) The store has several single-sheet, common glass windows. Will condensation form on the inside of the windows?

HVAC–9

8000 ft^3/min of 80°F dry bulb, 50% relative humidity air at 14.7 psia are cooled to a saturated 57°F condition in an air washer. 42°F chilled water for the spray is produced by an R-12 refrigeration system. The chilled water is heated to 55°F by the warm air before it is returned to the refrigerator.

(a) How much moisture (in lbm/hr) is removed from or added to the air in the washer?

(b) What is the tonnage of the refrigeration effect?

(c) What is the flow rate (in gpm) of the chilled water?

HVAC–10

A guinea pig breeder keeps 3000 pigs in stock for supply to pet stores around the country. The average pig weight is 3 lbm. The cage room is supplied with air at 82°F (dry bulb) and 85% relative humidity by an evaporative cooler. (There is no recirculation.) Air leaves the evaporative cooler at 140 ft/min. Air is removed from the cage room when it reaches 87°F. The cage room experiences the following loads.

equipment sensible heating	70,000 BTU/hr
lighting sensible heating	15,000 BTU/hr
guinea pigs	
sensible (per pound of pig)	4.5 BTU/lbm
latent (per pound of pig)	2.0 BTU/lbm

(a) What air flow (in ft^3/min) through the cage room is required?

(b) What is the humidity ratio of the air leaving the cage room?

(c) What is the area of the evaporative cooler supply duct?

HVAC–11

20,000 ft^3/hr (based on dry air at 60°F and 14.7 psia) of air are compressed to 100 psia and 500°F. The air is subsequently cooled to 75°F in a constant volume process that also saturates it. The compressor suction conditions are 13.4 psia, 85°F dry bulb, and 80% relative humidity.

(a) How much water (in lbm/hr) must be supplied or removed to saturate the cooled air?

(b) What is the specific humidity of the air entering the cooling process?

(c) What is the specific humidity of the air leaving the cooling process?

HVAC–12

A chemistry laboratory contains two fume hoods, each rated at 1400 ft^3/min. To ensure that a positive pressure exists in the room, 10% more air is supplied to the room than is withdrawn by the hood and room air return. This 10% is lost through leakage. 65% of the air removed by the fume hoods is supplied by the hoods' auxiliary duct work. The remaining 35% is supplied by the main room supply system. Assume all entering air is at the same condition. All air leaves the room at 72°F and 50% relative humidity. The air temperature increases 16°F between the times it enters the room and it is extracted by the fume hood exhausts. In addition, the following is known about the room.

occupancy	1 seated person per fume hood
room area	500 ft^2
lights heat gain	3 $watts/ft^2$
hot plates	18 $watts/ft^2$

(a) What is the heat gain in the room?

(b) What is the required volume of supply air?

(c) What are the conditions (dry bulb temperature and relative humidity) of the entering supply air?

(d) What are the volumes of the auxiliary fume hood supply, fume hood exhaust, and room air return? Draw a diagram, and label all entering and leaving air flows.

HVAC–13

The moisture content of an air stream is reduced by passing the air through a drying bed of silica gel. The drying bed is reactivated by heating the silica gel to drive off the moisture absorbed. One hour of reactivation is required for every five hours of drying (i.e., a 6-hr cycle). The following parameters describe this batch drying process.

air flow	15,000 ft^3/min
entering conditions	
dry bulb temperature	60°F
wet bulb	55°F
leaving conditions	
moisture content	25 grains/lbm air
silica gel	
specific heat	0.21 BTU/lbm-°F
heat of absorption	225 BTU/lbm vapor
moisture content of gel by weight	
just before reactivation	25%
just after reactivation	5%
latent heat of vaporization for water	1045 BTU/lbm

During the reactivation process, the gel drying beds are heated to 160°F and cooled back to 85°F by circulating cooler air through the beds. The cool air enters at 65°F and leaves at 75°F.

(a) What is the dry bulb temperature of the treated air leaving the dehumidifier?

(b) How much (in lbm) gel is required for a 6-hr cycle?

(c) How much (in ft^3/min) air is required to cool the reactivated beds back to 85°F in 1 hr?

(d) How much moisture is removed from the drying beds during reactivation?

HVAC–14

Two streams of moist air enter a mixing chamber and exit at 14.7 psia.

stream 1	
flow rate	1200 ft^3/min
relative humidity	100%
dry bulb temperature	60°F
pressure	14.7 psia
stream 2	
flow rate	1000 ft^3/min
relative humidity	100%
dry bulb temperature	110°F
pressure	14.7 psia

(a) What is the dry bulb temperature of the mixture?

(b) How much liquid water is withdrawn from the mixing chamber?

HVAC–15

Twelve people work in a small R&D building. Heated air from the room is withdrawn, mixed with make-up air, conditioned, mixed with bypassed air, and returned to the building.

outside design conditions	
dry bulb temperature	96°F
relative humidity	60%
indoor design conditions	
dry bulb temperature	73°F
wet bulb temperature	57°F
building volume	75,000 ft^3
occupancy	12 people seated
external heat gain	14,000 BTU/hr
internal heat loads	
lighting (sensible)	15 kW
motors	18,000 BTU/hr
air changes per hour	4
make-up air for exfiltration	10% of total supply
chilled water temperature rise	10°F

(a) What is the temperature of the air entering the room?

(b) What is the mass of air entering the coils?

(c) What is the tonnage of the air conditioner?

(d) What is the apparatus dew point?

(e) What amount (in gpm) of chilled water is required in the air conditioner?

(f) If the chilled water is circulated by a pump with an efficiency of 74% and operating against a 32-ft head, what is the brake horsepower of the pump?

HVAC–16

2750 gpm of 120°F water are cooled to 85°F in an evaporative, counterflow air cooling tower. Air enters the tower at 72°F dry bulb and 40% relative humidity and leaves at 100°F and 85% relative humidity. The windage and blow-down losses are 35% of the evaporative loss. Make-up water is available at 53°F. Atmospheric pressure is 14.7 psia.

(a) What is the air flow (in ft^3/min) through the cooling tower?

(b) How much (in gpm) make-up water is required?

HVAC–17

A school supply storage room has dimensions of 30 ft × 60 ft × 10 ft. During the winter, the outside conditions are 0°F and 30% relative humidity. The room is maintained at 68°F dry bulb and 25% relative humidity during the winter. 3% of the heated air in the room is lost each minute to leakage and is replaced by outside (untreated) air. 55°F water is evaporated in an air conditioner/heater to maintain the humidity as required. Slab losses are insignificant.

What will be the energy savings (in BTU/hr) if the room conditions are changed to 65°F dry bulb and 25% relative humidity?

HVAC–18

The cooling load for a real estate office in Albuquerque, New Mexico, is being calculated at 4:00 p.m. in July. The office is on the second floor of a large three-story building, and its windows face north. The following parameters describe the room.

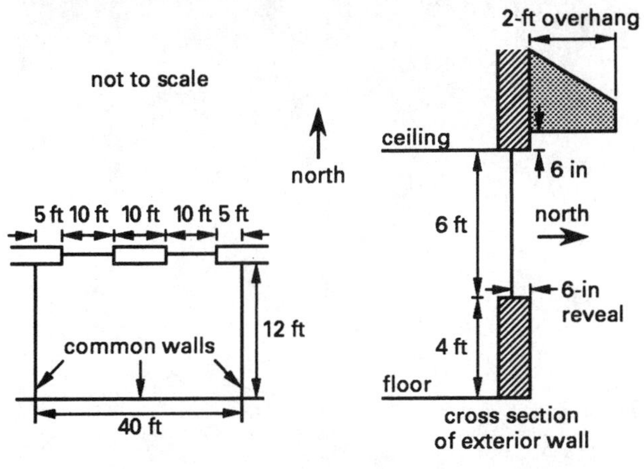

inside design conditions
- dry bulb temperature: 75°F
- relative humidity: 50%

outside design conditions at 4:00 p.m. in July
- dry bulb temperature: 96°F
- relative humidity: 60%

wall construction
- $\frac{5}{8}$-in gypsum inside
- $3\frac{1}{2}$-in batt insulation
- 4-in dark-colored brick face outside

windows
- size = 6 ft × 10 ft
- shading constant = 0.93
- $U = 0.33$ BTU/hr-ft^2-°F
- overhang: 2-ft horizontal projection (over windows only)

ceiling height: 10 ft

occupancy (all seated): 3 men

ventilation (per person): 25 ft^3/min

internal heat sources
- room lighting: 6 fixtures, four 40-watt tubes each

air conditioner
- chilled water temperature: 42°F
- water temperature rise: 15°F
- discharged air temperature: 57°F

(a) What is the total cooling load (in BTU/hr) on the air conditioner?

(b) How much (in gpm) cooling water is required?

(c) How much (in ft^3/min) air should be supplied to the room?

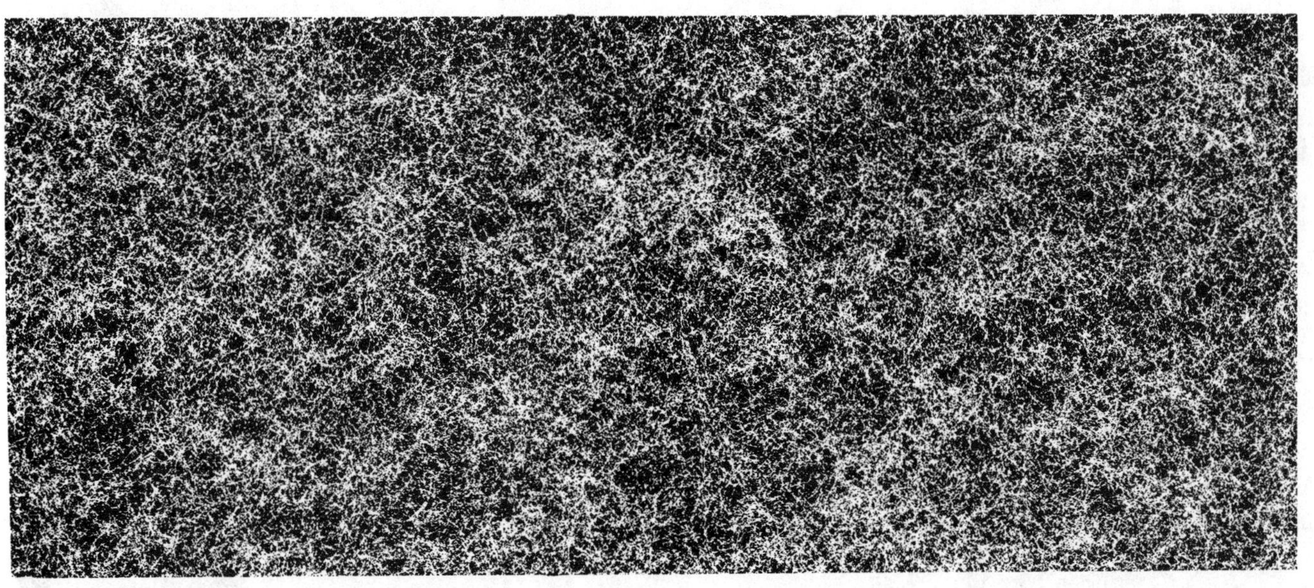

9 MECHANICS OF MATERIALS

MECHANICS OF MATERIALS–1

A steel bracket used as an anchor in race car traction tests is welded ($\frac{3}{8}$-in fillet weld) to a steel support as shown. The welding rod (electrode) designation is E6010. The load is steady, and a factor of safety of 2.5 is required. What is the maximum traction force, P, that can be supported at the force angle shown?

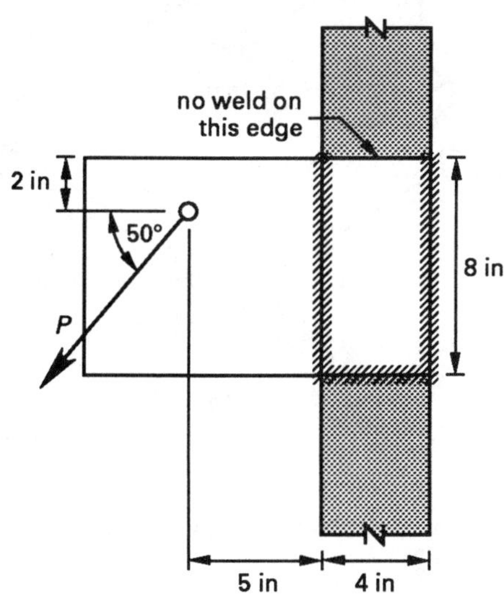

MECHANICS OF MATERIALS–2

A 12-in wide section of lap rivet joint in an aluminum sheet metal connection must support a tensile load of 2000 lbf. Both pieces of aluminum in the joint have a thickness of 0.057 in. The rivets are originally $\frac{1}{8}$ in in diameter, but are to be installed in holes produced with a #29 drill bit prior to heading. (Assume the rivets expand to completely fill the holes.) Cold-driven rivet strengths are higher than the strengths of the parent sheets, but this increase is to be disregarded. Similarly, the slight effect of eccentricity along the line of action is to be disregarded. Design an overlay (lap) rivet joint. The following material properties should be used.

$$S_{yt} = 35 \text{ ksi}$$
$$S_{yc} = 32 \text{ ksi}$$
$$S_{ys} = 22 \text{ ksi}$$
$$S_{yp} = 62 \text{ ksi (bearing)}$$
$$E = 10,000 \text{ ksi}$$
$$G = 3.8 \text{ ksi}$$
$$\text{Poisson ratio} = 0.33$$
$$\text{factor of safety} = 2.5 \text{ (based on yield properties)}$$

MECHANICS OF MATERIALS–3

The thrust of a small rocket engine is determined by observing the deflection of the test bed caused by that thrust. The test bed consists of a solid, rigid (but massless) platform supported at each corner by a 36-in high, 2-in wide leaf spring. Each leaf spring is constructed of ASTM 1050, #5 gage steel ($E = 2.9 \times 10^7$ psi). Both ends of each spring are considered to be rigid (i.e., fixed, built-in). A deflection of 2.0 in is observed.

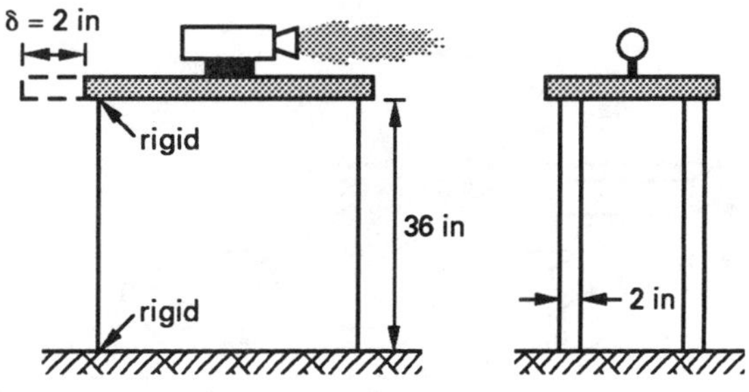

(a) What is the rocket engine thrust?

(b) What is the maximum bending stress in the springs?

MECHANICS OF MATERIALS–4

Two 90° rosette strain gages (yielding four simultaneous readings) are properly installed on a material experiencing stresses in unknown directions. The material has a modulus of elasticity of 300,000 MPa. The Poisson ratio is 0.33. The unit strains detected by the strain gages are

$\epsilon_1 = 0.0005$

$\epsilon_2 = -0.00245$ (negative indicates compression)

$\epsilon_3 = -0.0028$ (negative indicates compression)

$\epsilon_4 = 0.00015$

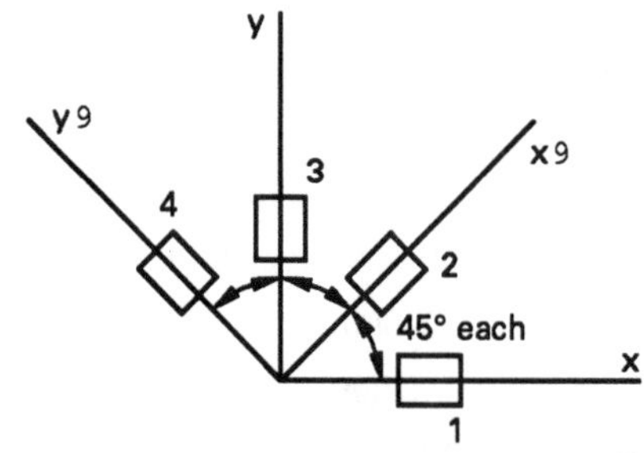

(a) What are the principal normal stresses?

(b) What is the principal shear stress?

MECHANICS OF MATERIALS–5

A rigid bar carrying a concentrated 120-kip load midway between its ends is supported by three square flexible steel ($E = 2.9 \times 10^7$ psi) beams. Each of the beams is simply supported at its two ends. The dimensions of the three support beams are

beam A:	3.0 in × 3.0 in
beam B:	3.25 in × 3.25 in
beam C:	3.5 in × 3.5 in

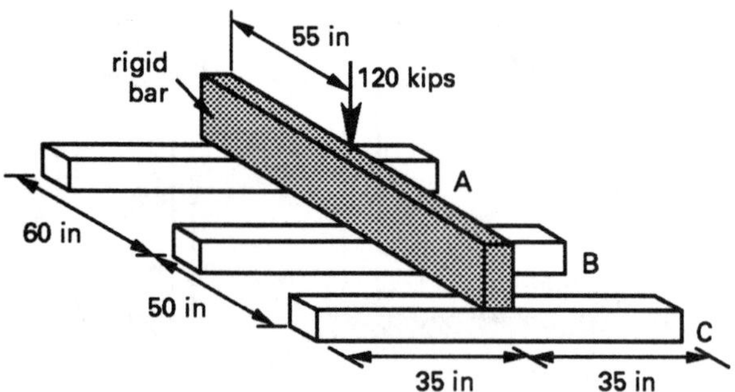

What are the deflections at the centers of beams A, B, and C?

MECHANICS OF MATERIALS–6

A column carrying a 12-kip load is constructed from round, structural steel tubing (2.375-in o.d., 2.067-in i.d.). The steel has a modulus of elasticity of 2.9×10^7 psi and a yield strength of 42,000 psi. The unbraced length is 11.5 ft. The column is fixed against rotation and translation at both ends of the unbraced length.

(a) Will this column fail under the 12-kip load?

(b) What is the factor of safety?

MECHANICS OF MATERIALS–7

A single concentrated load of 175 lbf is supported by a beam (2.25 in × 3.25 in) which, in turn, is supported by four 1.125-in diameter torsion bars. The torsion bars are fixed against rotation at each end (i.e., at the beam and at the distant end). The modulus of elasticity of the torsion bars and beam is 3.0×10^7 psi. The shear modulus for the torsion bars and beam is 1.1×10^7 psi.

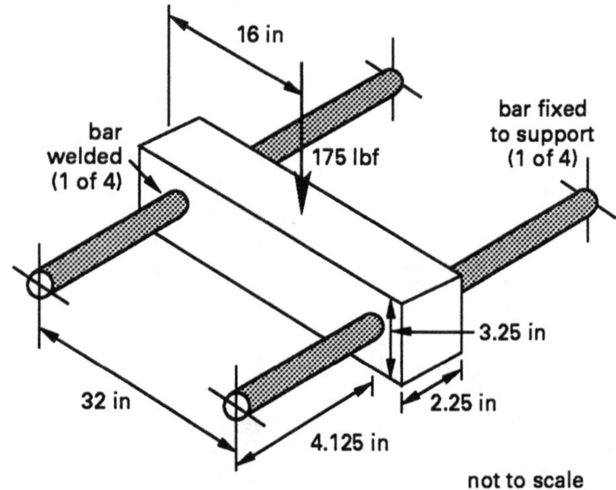

(a) What is the torque acting at each fixed torsion bar end?

(b) What is the maximum bending moment in the beam?

(c) What is the deflection at the point of loading?

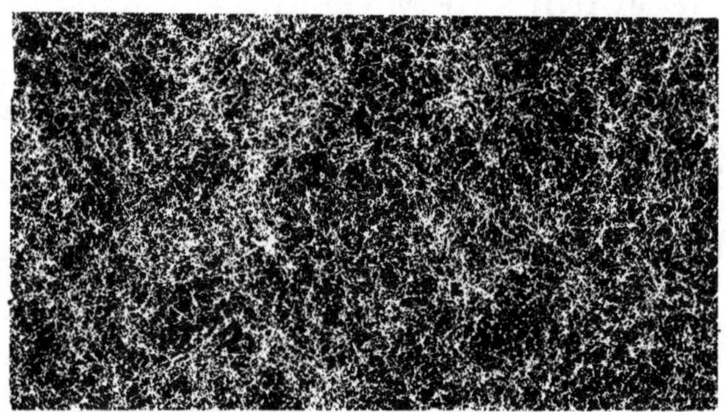

10 FAILURE THEORY

FAILURE THEORY–1

An interior airlock handle of an interstellar spacecraft is considered to be loaded in fully reversed bending and was originally designed for an infinite lifetime at 200% of the actual stress experienced. The handle was constructed of cold-rolled plain-carbon steel with the dimensions and properties given.

	cold-rolled	annealed
S_{ut}	80 ksi	60 ksi
S_{yt}	65 ksi	40 ksi
S_e	40 ksi	30 ksi

During repairs following a hostile alien attack, the handle was accidentally damaged by a plasma arc cutting torch, reducing the stress area of the bar as shown. Estimate the number of airlock cycles the bar will withstand before failure. State all assumptions.

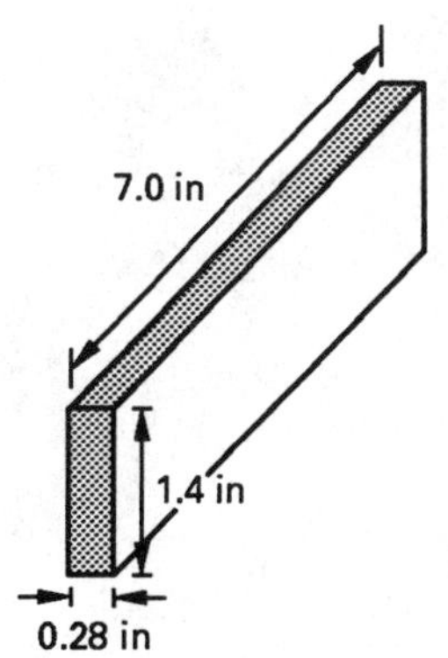

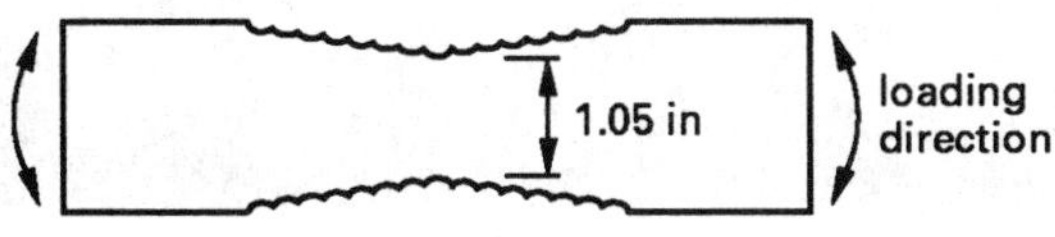

FAILURE THEORY–2

A forged steel connecting rod cap in a high-pressure, reciprocating air compressor is expected to experience multiple repetitions of 24,000 lbf compression and 7000 lbf tension. The rod end cap will be attached to the rod with two $\frac{3}{8}$-in 24 UNF grade 5 steel bolts, preloaded to 80% of the bolt material proof strength. The effective stress area under the bolt heads is limited by the cap lands to 0.32 in^2 for each bolt. All of the threads in the cap bolt engage the rod. A stress concentration factor of 2.65 should be used for the bolt threads. The bolt properties are

$$S_{ut} = 120 \text{ ksi}$$
$$S_{yt} = 80 \text{ ksi (proof strength)}$$
$$S_e = 40 \text{ ksi}$$

(a) What is the alternating stress in the bolts?

(b) Is the design adequate?

(c) What are the considerations in choosing between steel and aluminum for connecting rod design?

FAILURE THEORY–3

In order to operate in a non-sparking submarine environment, a high-pressure air line (2.25-in i.d., 0.20-in wall thickness) was manufactured from a copper alloy (S_{yt} = 16 ksi, S_{yc} = 16 ksi, and S_{ut} = 42 ksi). The line carries high-pressure air at 750 psig to the bow water (ballast) tanks. During a depth charge attack, the air line experiences a torsional stress. Neglecting instability and longitudinal stress, what is the maximum torque (in in-lbf) that the air line can withstand without yielding?

FAILURE THEORY–4

A solid round bar, 32 in long, is constructed from ductile steel (S_{yt} = 60,000 psi, $G = 11.5 \times 10^6$ psi). The bar is fixed at one end and is free to twist at the other.

The bar is used as a torsion bar spring having a spring constant of 3600 in-lbf/radian.

(a) What is the required diameter of the bar?

(b) What is the maximum torsional stress that the bar can withstand?

(c) What is the maximum allowable torque?

(d) What is the stress in the bar if an 8° twist is experienced?

FAILURE THEORY–5

A long pipe (3.5-in mean radius, 0.25-in wall thickness) is constructed from steel (S_{yt} = 50 ksi). The pipe carries high-pressure, room-temperature carbon dioxide at 800 psig. Due to failure in several poorly installed pipe hangers, a 190,000 in-lbf torque is also experienced. Disregarding longitudinal stress, what is the factor of safety in this loading condition?

FAILURE THEORY–6

The main member supporting a ferris wheel bench is being designed from steel ($S_e = 0.33\ S_{ut}$). The member is 48 in long and has a 1.5-in × 1.5-in (solid) square cross section. The ends are effectively built-in (i.e., fixed) as shown below. The member can be considered to be loaded repeatedly by a completely reversing 1500-lbf force applied at member midspan, which includes the weight of the bench and maximum passenger load. Disregarding any notch sensitivity or stress concentration effects at the loading point and using a factor of safety of 1.5, what should be the ultimate tensile strength of the steel?

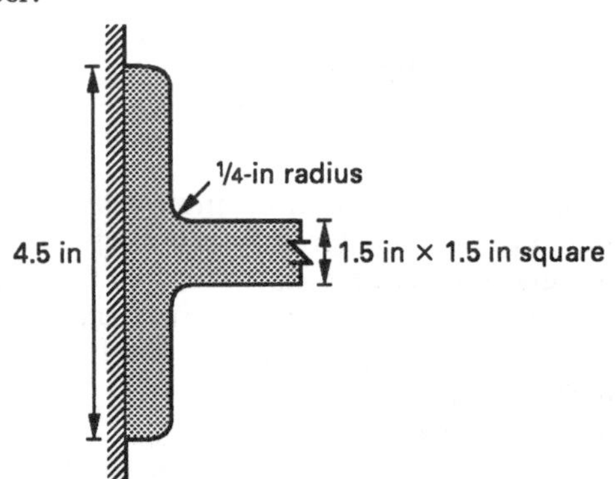

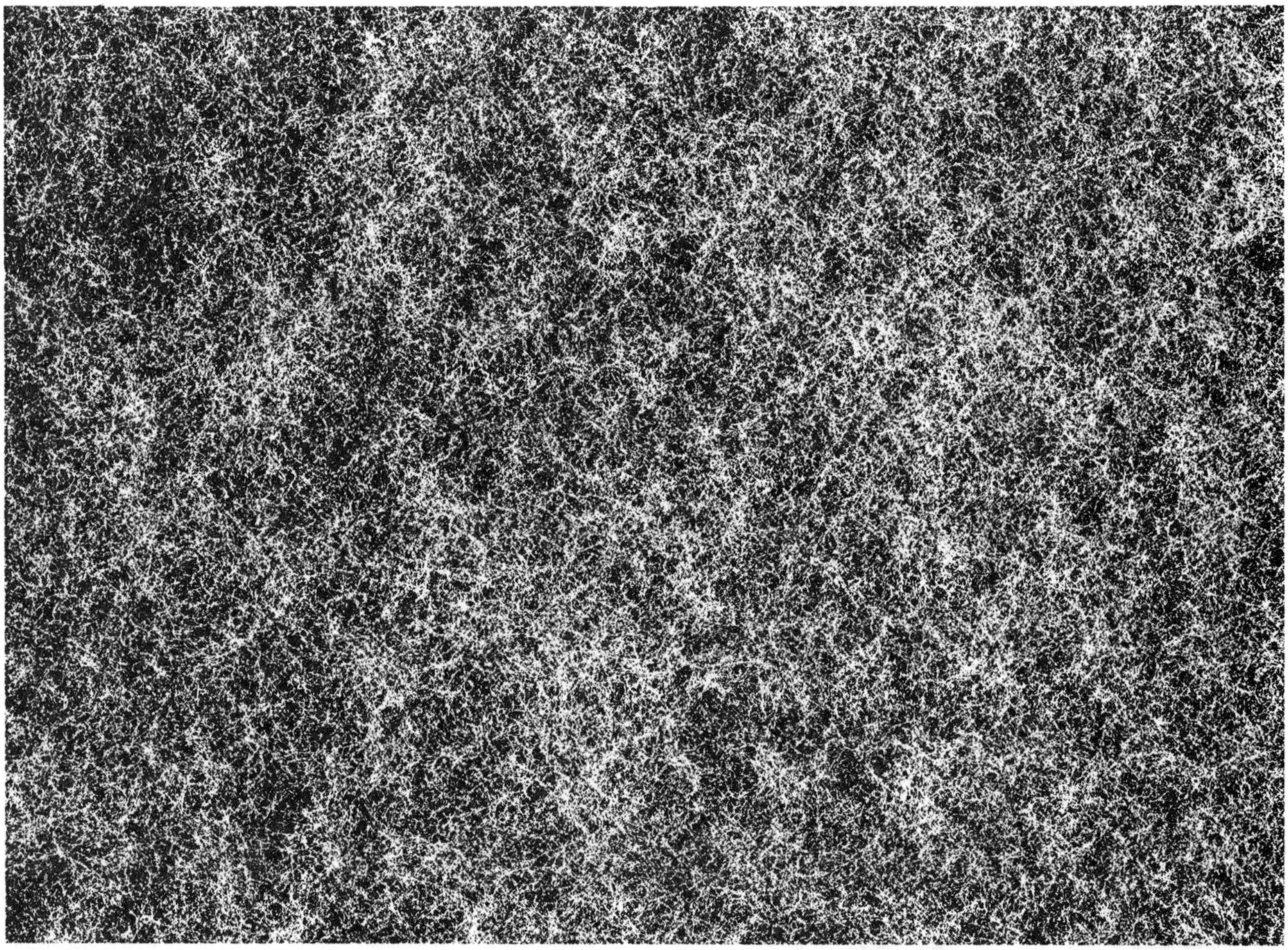

11 MACHINE DESIGN

MACHINE DESIGN–1

A $1\frac{1}{2}$-in thick, 12-in o.d. aluminum disk is shrink-fitted to a 2-in steel shaft with a diametral interference of 0.0045 in. What rotational speed (in rpm) will reduce the interference to zero?

MACHINE DESIGN–2

The longitudinal axes of two parallel precision shafts are 8.5 in apart. Each shaft carries a 20° full-depth gear, and the two gears are in mesh, providing a nominal 2.5:1 speed reduction. Determine the diametral pitch, pitch diameters, and number of teeth for each of the two gears such that the contact ratio is between 1.5 and 1.6.

MACHINE DESIGN–3

A bronze coupling sleeve (nominal 7-in o.d., 6-in i.d., 16 in long) connects two solid steel shafts (nominal 6-in o.d.). The sleeve is not keyed, splined, or pinned. The two shafts turn at 1800 rpm, and 3000 ft-lbf of torque are transmitted through an interference fit. The coefficient of static friction between the bronze coupling sleeve and the steel shafts is 0.16. The shafts operate in a 190°F environment, but the coupling sleeve is press-fitted when sleeve and shafts are at 70°F. Use the following data to determine what the initial (at 70°F) interference should be.

	steel	bronze
coefficient of thermal expansion	6.5×10^{-6} 1/°F	10.2×10^{-6} 1/°F
modulus of elasticity	2.9×10^{7} psi	1.6×10^{7} psi
shear modulus	1.1×10^{7} psi	6.0×10^{6} psi
Poisson ratio	0.30	0.33

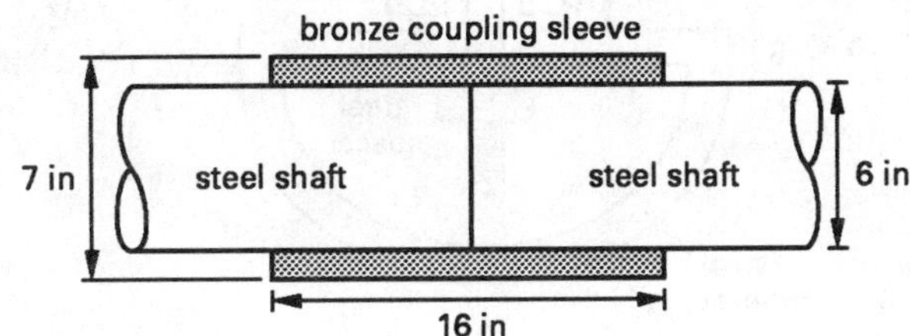

MACHINE DESIGN–4

A spur gear in a taffy-pulling apparatus is currently manufactured from stainless steel. It has been proposed that a nylon-based gear of the same design be used to replace the more expensive stainless steel gear. The data below describes the parameters governing the replacement decision. Should the nylon gear be considered?

gear quality	commercial
allowable nylon gear stress	14,000 psi
face width (all gears in mesh)	0.35 in
tangential load on gear tooth	60 lbf
contact ratio	1.69
pitch diameter	1.21 in
pressure angle	20° (full depth)
number of teeth	34
stress concentration factor	1.45
speed	very slow
temperature	very low; controlled

MACHINE DESIGN–5

A grade 5, 1-in UNF-12 steel bolt holds a steel bracket as shown. The bolt is initially preloaded to an internal tensile force of 14,000 lbf. There are no threads between the nut and bolt head. The bracket supports a load that varies continuously from 6000 lbf to 12,000 lbf. Disregard the bracket stresses and the stiffness/flexibility of the plate member to which the bolt bracket is secured. Find the factor of safety for the bolt.

spacer elastic modulus	2.9×10^7 psi
bolt elastic modulus	2.9×10^7 psi
bolt S_e	70,000 psi
bolt S_{yt}	140,000 psi
k	3.7 (stress concentration factor for bolt threads)

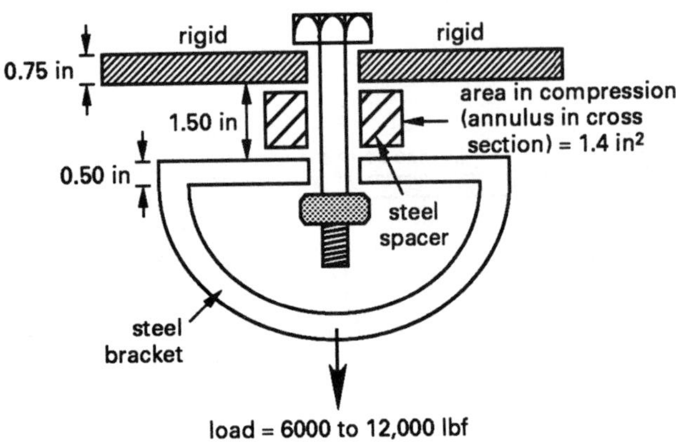

MACHINE DESIGN–6

A 50-hp motor runs at a synchronous speed of 1800 rpm. The motor drives a set of gears that provides a nominal speed reduction of 2:1 between parallel shafts mounted 9.0 in apart. Both gears are steel, have a 20° pressure angle (full-depth), and are class 1 (well-cut, precision, high-quality, etc.). The yield stress in bending for both gears is 38,000 psi, and a factor of safety of 2.0 should be used. The pinion (mounted on the motor armature shaft) has 24 teeth. For each gear, find the

(a) diametral pitch

(b) gear pitch line diameter

(c) face width

MACHINE DESIGN–7

A mechanism to raise and lower an 800-lbm interior check gate in a high-security prison is being designed. The gate is suspended at two points by two high-test aircraft cables. Each cable carries half of the gate load. Each cable is wrapped around a 10-in diameter pulley mounted on an AISI 1040 steel shaft (S_{yt} = 60,000 psi, $E = 2.9 \times 10^7$ psi). The gate is raised and lowered by electric motors working in unison through additional cabling and 14-in diameter pulleys located at each shaft end. The shaft supporting the gate is continuous, and the effects of all keyways, holes, splines, etc., are to be neglected, as is the effect of fatigue. Using a factor of safety of 2.0, determine the required shaft diameter.

10 in
8 in
30 in
8 in
10 in
14-in diameter pulley (typical)
bearing
actuating motor cable
10-in diameter pulley
bearing
actuating motor cable
gate mass = 800 lbm

MACHINE DESIGN–8

It is desired to strengthen an existing short steel pipe (8.000-in o.d., 0.450-in wall thickness) in order to increase the pressure the pipe can withstand. A steel jacket (7.990-in i.d., 0.30-in wall thickness) is heated and slipped over the pipe. After assembly and cooling, the pipe is pressurized to 750 psig internally. For both the pipe and jacket steel, E (the modulus of elasticity) is 2.9×10^7 psi, and μ (the Poisson ratio) is 0.30. There is no longitudinal stress. What are the maximum tangential stresses in the pipe and the jacket?

MACHINE DESIGN–9

2000 lbm of sprung automobile mass are supported by two helical coil springs. (The sprung mass can be considered to be independent of other car mass, and damping is to be neglected. A bar linkage keeps the two springs in phase.) For the personal comfort of passengers, a natural vertical oscillation frequency for the sprung mass of 1.25 Hz is desired. Also, manufacturing constraints make it necessary to keep the spring index as close to 8.0 as possible. The torsional yield stress for the spring material is 60,000 psi, the shear modulus is 11.5×10^6 psi, and a factor of safety of 1.75 is to be used. The spring coils are solid at a 7-in deflection. For the springs, find the theoretical (a) mean coil diameter, (b) wire diameter, and (c) number of active coils.

MACHINE DESIGN–10

A hanging hook designed for vertical loads in a parts warehouse is constructed of cold-drawn steel, 1.250 in in diameter. The maximum allowable bending stress for the hook material is 26,000 psi; the maximum allowable shear stress is 14,000 psi. The base of the hook is rigidly attached (fixed) to the wall. Through a misunderstanding, the hook is used to anchor a horizontal (static) load, assumed to be applied at the tip of the hook. Neglecting any stress concentration at the elbow, what is the maximum allowable force that can be applied at the tip?

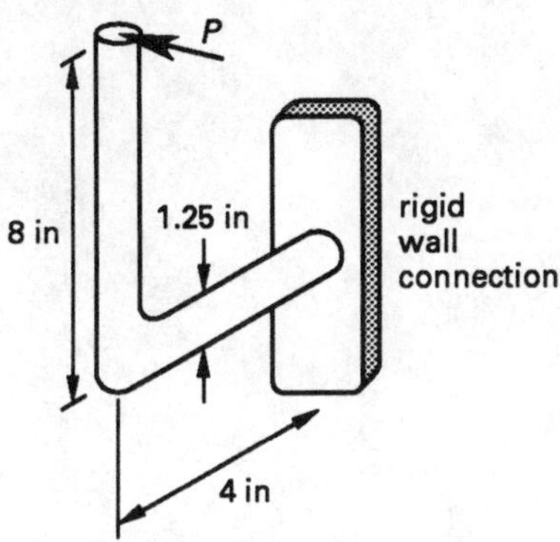

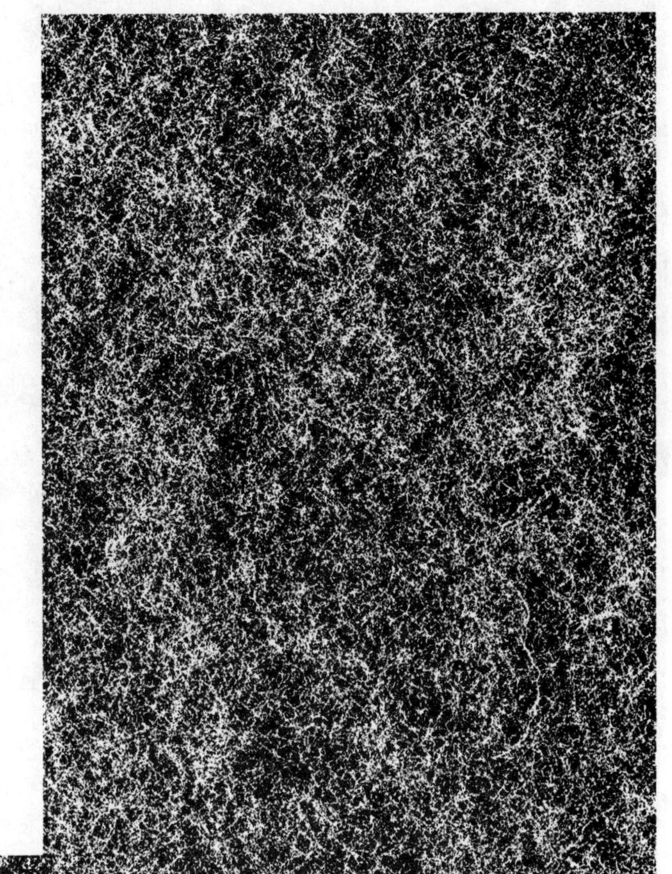

12 DYNAMICS

DYNAMICS–1

A variable speed reducer in a grain combine (where the engine runs constantly at an efficient 3000 rpm) is constructed as shown below. Gear B is an idler.

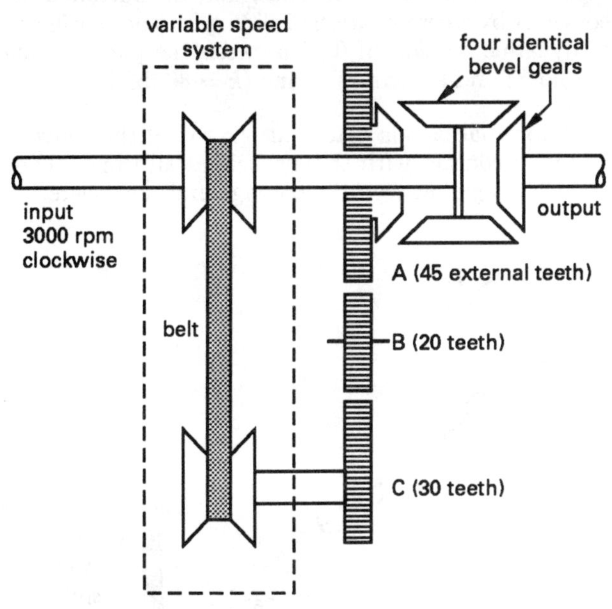

What are the output speed and direction if the variable speed system is set for a 40% speed reduction?

DYNAMICS–2

A motor (rotational mass moment of intertia of 25 lbf-sec^2-in) drives a load (rotational mass moment of inertia of 15 lbf-sec^2-in) through a set of speed-increasing gears. Neglecting the mass of the gears and gear tooth stiffnesses, determine the torsional natural frequency of the motor drive system. Use $G = 11.5 \times 10^6$ psi for the steel shafts.

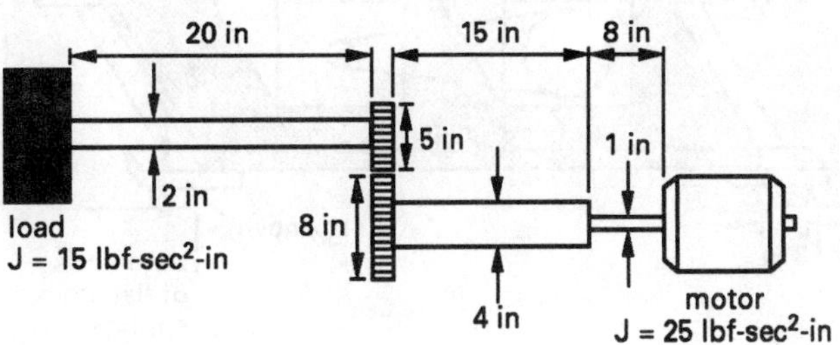

DYNAMICS–3

The rear suspension on a pickup truck used for pizza deliveries consists (essentially) of two coil springs between a 2-wheel axle and frame. The truck bed mass tributary to the springs is 550 lbm. The springs have six active coils, are wound on a 6-in nominal (mean) diameter, and are constructed from 0.84-in diameter steel wire (shear modulus = 11.5×10^6 psi, S_{ut} = 100 ksi). The stiffness of the tires, when properly inflated, is taken as 2200 lbf/in for each of the two tires.

Pizzas are kept warm in an 800-lbm oven. During installation, the oven is accidentally dropped 8 in onto the bed. The oven does not rebound from the bed.

(a) How far will the bed deflect due to the impact? (Disregard any structural damage.)

(b) How far will the springs (alone) deflect?

(c) What is the maximum deceleration (in gravities) experienced by the oven?

(d) Will the springs experience a stress in excess of their yield strength due to the oven impact load?

DYNAMICS–4

A planetary gear set has a fixed internal ring gear with 119 standard 20° spur teeth. There are two planets. The sun gear (the input) turns in the same direction and eight times as fast as the planet carrier (the output). How many teeth are on the sun and planet gears?

DYNAMICS–5

A small motor-driven compressor is mounted on a rigid metal plate, which in turn, is supported on four fixed-end leaf springs. The motor turns at 1750 rpm, and the compressor turns at 583 rpm. The mass of the motor, compressor, and plate totals 47 lbm. (Consider the mass to be evenly distributed.) The springs are constructed of steel ($E = 2.9 \times 10^7$ psi), 3 in long (clear span) and 1 in wide. Both ends of each spring are welded (i.e., the springs have fixed ends).

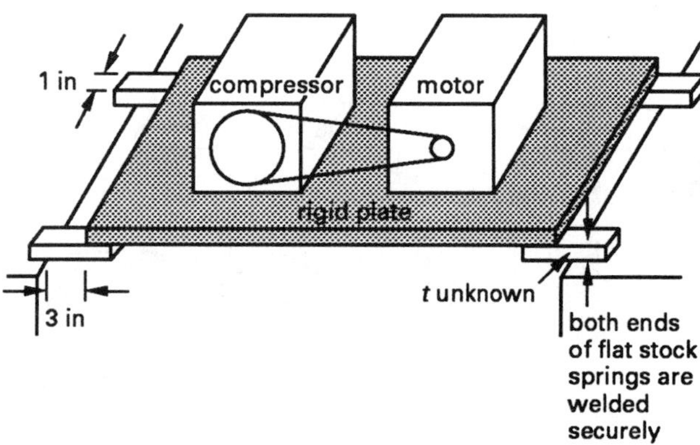

(a) What thickness should the springs be for the plate to achieve a natural frequency of 14 Hz?

(b) What is the maximum natural frequency that can be allowed such that damping of the motor's and compressor's vibrations will be effective?

DYNAMICS–6

An Everglades Glider swamp boat is propelled along the surface of the water by a 50-hp motor-driven propeller located behind the driver. The motor and propeller are connected by a 2.0-in diameter, 14-in long steel shaft ($G = 11.5 \times 10^6$ psi) in a direct-drive arrangement. The motor has an effective mass moment of inertia of 100 lbm-in^2, and turns at 2400 rpm. The propeller has a mass of 18 lbm and an effective radius of gyration of 2.75 ft, and turns at 2400 rpm. The driver accidentally runs into the branches of a tree, and the propeller is immediately stopped. (The propeller does not snap.) Simultaneously, the driver is knocked from the seat, releases the dead-man switch, and the motor instantly and simultaneously shuts off. What is the maximum shear stress in the shaft between the motor and propeller?

DYNAMICS–7

A steel flywheel (12 in in diameter and 1.25 in thick) is mounted on a 1.25-in diameter steel shaft, exactly midway between the two self-aligning bearings 18 in apart supporting the shaft.

(a) What is the percentage effect on the critical speed of increasing the span between bearings by 15%?

(b) What is the percentage effect on the critical speed of increasing the shaft diameter by 15%?

(c) What is the percentage effect on the critical speed of increasing the flyweel diameter by 15%?

(d) What is the percentage effect on the critical speed of increasing the flywheel thickness by 15%?

DYNAMICS–8

An unpowered roller conveyor in an air freight company is constructed on a 35° incline. Each roller is 2 in in diameter and has a mass of 10 lbm. The center-to-center spacing of the rollers is 3.0 in. Rollers have no rotational friction and are initially at rest.

Packages are slowed to a standstill at the bottom of the conveyor by a combination of sliding friction (coefficient of dynamic friction of 0.15 between the packages and the floor) and a bumper spring ($k = 80$ lbf/in).

A 40-lbm package is placed at the top of the conveyor and is in contact with three rollers at all times. There is no slippage between the package and the rollers.

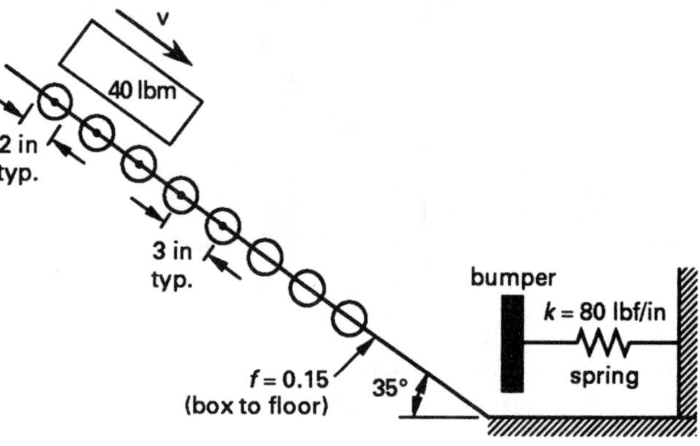

(a) Assuming that the inclined part of the conveyor is infinite in length, what would be the maximum velocity (parallel to the conveyor surface) achieved by the 40-lbm package?

(b) Assuming that the package comes off the conveyor at 80 in/sec, what is the maximum package deceleration (in gravities)?

(c) Assuming that the package comes off the conveyor at 80 in/sec, what value of the spring constant (k, in lbf/in) will limit the deceleration to 2 gravities?

DYNAMICS–9

10.0 ft-lbf of torque are transmitted to a 3-in pulley between two pulleys by way of a flat belt. The larger pulley is 12 in in diameter. The coefficient of friction between the belt and pulley is 0.35. To maintain proper tension in the belt, the two pulleys are forced apart by a spring ($k = 65$ lbf/in). While running, the belt incline is 30° (interior wedge angle of 60°).

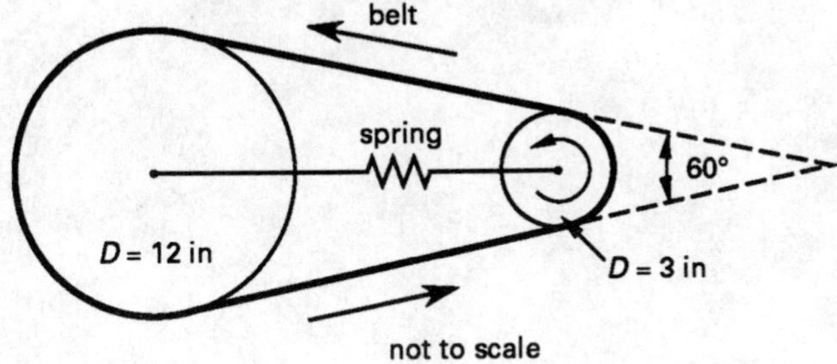

(a) What are the belt tensions?

(b) What is the deflection of the spring under load?

DYNAMICS–10

A thin-walled, low-carbon steel tube (0.20-in o.d., 0.025-in wall thickness, 35 in long, modulus of elasticity = 2.9×10^7 psi) is being considered as a medium-band antenna on an orbiting satellite. In operation, the antenna would be fixed at one end and free at the other end. There is some concern that on-board mechanical devices may set up resonance at the antenna's second modal frequency.

(a) What is the second modal frequency for the antenna?

(b) Sketch the first three modal shapes, and identify the nodes.

DYNAMICS–11

A manufacturer of carrying cases for portable computers claims its cases can withstand acceleration and deceleration of 8 gravities without damage. (No claim is made, however, about damage to the computer.) The cases are tested by filling them with lead shot until a total mass (case and lead) of 1300 lbm is achieved. The filled case is subsequently dropped from a height of 2.0 ft onto the floor, from which it does not rebound. For the purpose of this analysis, assume the force of the fall is carried by a single W6 × 25 steel beam floor joist with the following properties.

supports	simple, both ends
length	24 ft
loading point	mid-span (12 ft from ends)
area	7.34 in^2
tensile yield strength	36 ksi
modulus of elasticity	2.9×10^7 psi
moment of inertia	53.4 in^4
section modulus	16.7 in^3
joist mass	negligible

The building landlord has received complaints from other tenants about unexplained crashes, sagging floors, and falling plaster. In addition, pictures are falling from the walls throughout the building. When the landlord found out what the manufacturer was doing, he expressed concern about potential damage to the building.

(a) Should the landlord be concerned about his building?

(b) Does the carrying case experience more than 8 gravities of deceleration?

DYNAMICS–12

A car traveling along a road encounters a section where the asphalt road is being repaved, and a 1.2-in layer of new paving has been placed normal to the car's direction of travel. The car's front wheel hits the 1.2-in step, and the car mass (body) jumps 0.6 in above the new road surface. (That is, the body overshoots the road surface by 0.6 in.) The car mass tributary to each wheel is 1200 lbm. Each wheel is isolated from the car mass by a coil-over shock absorber. The spring rate of the spring is 90 lbf/in.

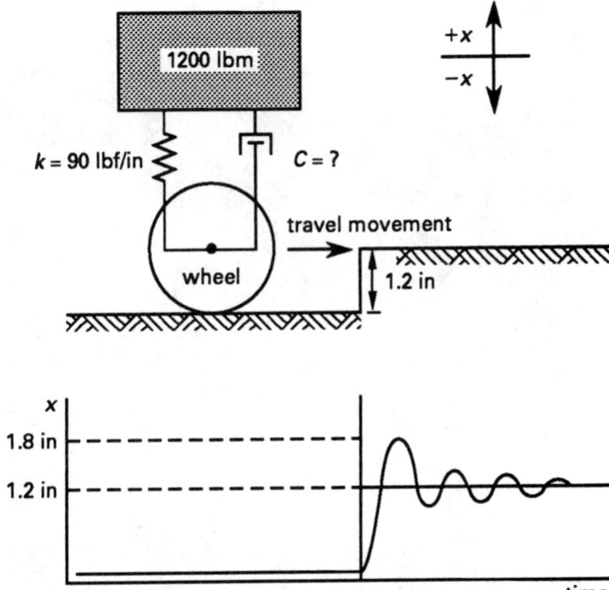

(a) What is the shock absorber's damping coefficient, C?

(b) How long will it take to reduce the amplitude of the oscillation to 20% of its maximum value?

(c) What is the natural frequency of oscillation?

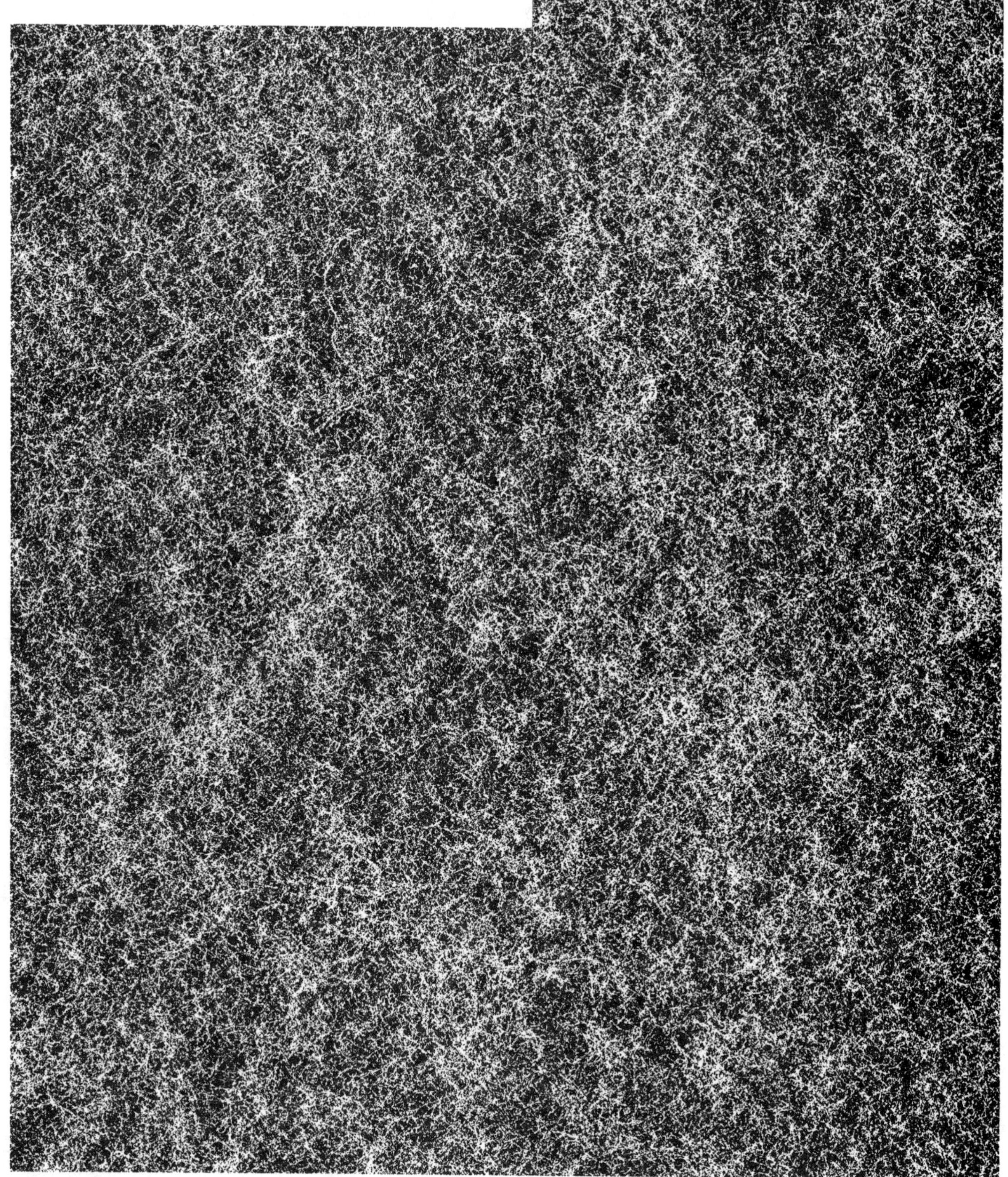

13 CONTROLS AND SYSTEMS

CONTROLS AND SYSTEMS–1

30.0 lbm/min of 180°F (nominal) water flow out of a large mixing tank at a constant rate. The tank receives 210°F saturated steam at a controlled rate of up to 15 lbm/hr. (The maximum rate applies to a fully-opened control valve.) The tank also receives 60°F water at a constant rate. The incoming water and steam experience ideal mixing, and the resultant temperature of the water in the tank is initially 180°F. The tank initially holds 600 lbm of water. Ignore all heat transfer to the surroundings. Assume the valve position changes instantaneously.

The temperature of the tank output is monitored by a controller, which controls the entering steam rate. Two types of controllers are available: on/off and proportional. Both units have sufficient capacity and duty cycle to handle the full range of steam flows (up to 15 lbm/min) expected. The proportional controller has the following characteristics.

temperature variation between fully opened and fully closed signal	12°F
acceptable temperature set point	160°F – 200°F

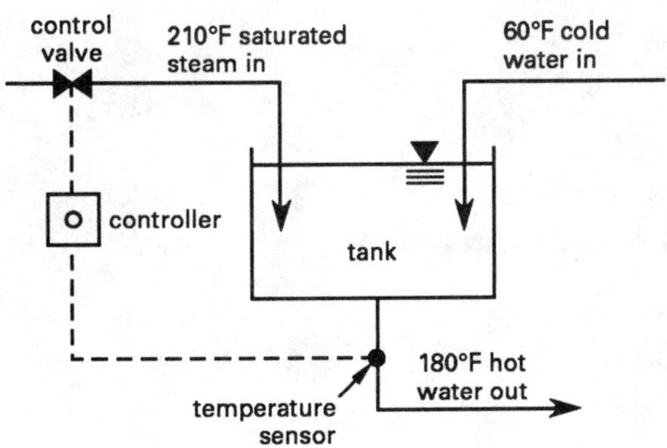

(a) If the on-off controller is selected, and if the mass of the water in the tank is constant, how long (in seconds) must the valve be fully opened each minute to keep the tank temperature at 180°F?

(b) If the steam valve jams fully open, and if there are no changes in the cold water input and mixed water output rates, how long will it take to heat the tank to 190°F?

(c) If 180°F water is desired, what is the proportional band for the proportional controller?

(d) If 180°F water is desired, what should be the set point?

CONTROLS AND SYSTEMS–2

You are in charge of a motorcycle assembly plant. Strikes at your assembly plant have limited the labor and material resources available to you. You can either produce a Turbo-Cycle (with an accompanying profit of $500), or you can produce a Punk-Moped (with an accompanying profit of $300). You would like to maximize your profit.

Unfortunately, it is not a simple matter of merely making all Turbo-Cycles, since certain parts are in short supply. (All parts can be used by either cycle.)

part	number available
grease fittings	1800
running lights	1600
fuses	4400

Due to differences in cycle design, the quantity needed of each of the scarce parts varies.

	number used in	
part	Turbo-Cycle	Punk-Moped
grease fittings	6	2
running lights	4	2
fuses	6	8

How many Turbo-Cycles and Punk-Mopeds should you have your assembly plant produce in order to maximize profit?

CONTROLS AND SYSTEMS–3

A control system feedback loop is modeled as shown.

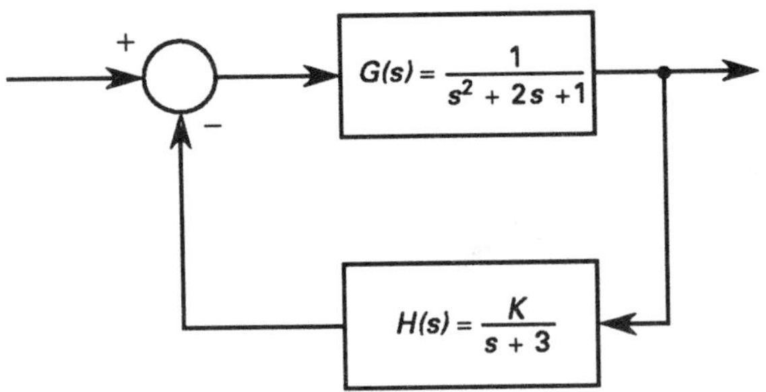

(a) What is the transfer function?

(b) Use the analytical or graphical method of your choice to determine the stability of this feedback loop.

ENGINEERING ECONOMIC ANALYSIS

1. Compare the present worths.

Batch:

$$D_j = \frac{25{,}000 - 3000}{5} = 4400$$

$$\begin{aligned} P(\mathrm{B}) &= -25{,}000 - (20{,}000)(P/A, 15\%, 5)(1 - 0.48) \\ &\quad + (3000)(P/F, 15\%, 5) \\ &\quad + (4400)(P/A, 15\%, 5)(0.48) \\ &\quad + (950{,}000)(0.10)(P/A, 15\%, 5)(1 - 0.48) \\ &= -25{,}000 - (20{,}000)(3.3522)(0.52) \\ &\quad + (3000)(0.4972) + (4400)(3.3522)(0.48) \\ &\quad + (950{,}000)(0.10)(3.3522)(0.52) \\ &= -25{,}000 - 34{,}863 + 1492 + 7080 + 165{,}599 \\ &= 114{,}308 \end{aligned}$$

This is positive, so ROR > 15%.

Continuous:

$$D_j = \frac{50{,}000 - 6000}{5} = 8800$$

$$\begin{aligned} P(\mathrm{C}) &= -50{,}000 - (18{,}000)(P/A, 15\%, 5)(1 - 0.48) \\ &\quad + (6000)(P/F, 15\%, 5) \\ &\quad + (8800)(P/A, 15\%, 5)(0.48) \\ &\quad + (1{,}000{,}000)(0.10)(P/A, 15\%, 5)(1 - 0.48) \\ &= -50{,}000 - (18{,}000)(3.3522)(0.52) \\ &\quad + (6000)(0.4972) + (8800)(3.3522)(0.48) \\ &\quad + (1{,}000{,}000)(0.10)(3.3522)(0.52) \\ &= -50{,}000 - 31{,}377 + 2983 + 14{,}160 + 174{,}314 \\ &= 110{,}080 \end{aligned}$$

This is positive, so ROR > 15%.

Since $P(\mathrm{B}) > P(\mathrm{C})$, choose the batch process.

(From a practical economic standpoint, these two alternatives are essentially identical. Other factors may be considered to complete the evaluation.)

2.

$$D_1 = \left(\frac{2}{5}\right)(10{,}000) = 4000$$

$$D_2 = \left(\frac{2}{5}\right)(10{,}000 - 4000) = 2400$$

$$D_3 = \left(\frac{2}{5}\right)(10{,}000 - 4000 - 2400) = 1440$$

$$\mathrm{BV}_3 = 10{,}000 - 4000 - 2400 - 1440 = 2160$$

Gain on the sale of the depreciated asset at the end of year 3 is

$$5500 - 2160 = 3340$$

The tax credit is

$$(0.03333)(10{,}000) = 333 \text{ at } t = 1$$

$$\begin{aligned} P &= -10{,}000 + (4000)(P/F, 15\%, 1)(0.40) \\ &\quad + (2400)(P/F, 15\%, 2)(0.40) \\ &\quad + (1440)(P/F, 15\%, 3)(0.40) \\ &\quad + (5500)(P/F, 15\%, 3) \\ &\quad - (3340)(P/F, 15\%, 3)(0.40) \\ &\quad + (333)(P/F, 15\%, 1) \\ &\quad - (500)(P/A, 15\%, 3)(1 - 0.40) \\ &= -10{,}000 + (4000)(0.8696)(0.40) \\ &\quad + (2400)(0.7561)(0.40) \\ &\quad + (1440)(0.6575)(0.40) + (5500)(0.6575) \\ &\quad - (3340)(0.6575)(0.40) + (333)(0.8696) \\ &\quad - (500)(2.2832)(0.60) \\ &= -10{,}000 + 1391 + 726 + 379 \\ &\quad + 3616 - 878 + 290 - 685 \\ &= \boxed{-5161} \end{aligned}$$

3. The previous three years are irrelevant. This should be evaluated as a four-year proposal.

$$D_1 = (10{,}000)(0.25) = 2500$$

$$D_2 = (10{,}000)(0.38) = 3800$$

$$D_3 = (10{,}000)(0.37) = 3700$$

At the end of three years, the book value is zero. The gain on the sale of a depreciated asset is 2000.

Inspector uses own car:

$$\begin{aligned} P(\mathrm{C}) &= -(0.30)(12{,}000)(P/A, 15\%, 4)(1 - 0.40) \\ &= -(0.30)(12{,}000)(2.8550)(0.60) \\ &= -6167 \end{aligned}$$

Buy a truck:

$$\begin{aligned}P(\mathrm{T}) &= -10{,}000 - (1800)(P/A, 15\%, 4)(1 - 0.40)\\&\quad + (2000)(P/F, 15\%, 4)\\&\quad + (2500)(P/F, 15\%, 1)(0.40)\\&\quad + (3800)(P/F, 15\%, 2)(0.40)\\&\quad + (3700)(P/F, 15\%, 3)(0.40)\\&\quad - (2000)(P/F, 15\%, 4)(0.40)\\&= -10{,}000 - (1800)(2.8550)(0.60)\\&\quad + (2000)(0.5718) + (2500)(0.8696)(0.40)\\&\quad + (3800)(0.7561)(0.40)\\&\quad + (3700)(0.6575)(0.40)\\&\quad - (2000)(0.5718)(0.40)\\&= -10{,}000 - 3083 + 1144 + 870\\&\quad + 1149 + 973 - 457\\&= -9404\end{aligned}$$

Since 9404 > 6167, continue paying the inspector for personal car use.

4. The interest rate is unknown. Work in thousands of dollars.

$$D_j = \frac{525 - 0}{15} = 35$$

$$\mathrm{BV}_7 = 525 - (7)(35) = 280$$

Gain on the sale of a depreciated asset is

$$700 - 280 = 420$$

$$\begin{aligned}P &= -525 + (700)(P/F, i\%, 7)\\&\quad + (35)(P/A, i\%, 7)(0.40)\\&\quad - (420)(P/F, i\%, 7)(0.40)\\&\quad - (25)(P/A, i\%, 7)(1 - 0.40)\\&\quad + (45)(P/A, i\%, 3)(1 - 0.40)\\&\quad + (75)(P/A, i\%, 7)(1 - 0.40)\\&\quad - (75)(P/A, i\%, 3)(1 - 0.40)\end{aligned}$$

This analysis neglects the benefit of a favorable capital gains tax rate on the gain above 525.

Simplify and combine terms.

$$\begin{aligned}P &= -525 + [700 - (420)(0.40)](P/F, i\%, 7)\\&\quad + [(35)(0.40) - (25)(0.60)\\&\quad + (75)(0.60)](P/A, i\%, 7)\\&\quad + [(45)(0.60) - (75)(0.60)](P/A, i\%, 3)\\&= -525 + (532)(P/F, i\%, 7) + (44)(P/A, i\%, 7)\\&\quad - (18)(P/A, i\%, 3)\end{aligned}$$

Try $i = 10\%$:

$$\begin{aligned}P &= -525 + (532)(0.5132) + (44)(4.8684)\\&\quad - (18)(2.4869)\\&= -82.5\end{aligned}$$

Try $i = 5\%$:

$$\begin{aligned}P &= -525 + (532)(0.7107) + (44)(5.7864)\\&\quad - (18)(2.7232)\\&= 58.7\end{aligned}$$

Use linear interpolation.

$$\mathrm{ROR} \approx 5\% + (10\% - 5\%)\left(\frac{58.7}{58.7 - (-82.5)}\right) = \boxed{7.08\%}$$

5. This is a replacement study, not an alternative comparison problem.

Since efficiency and production capacity are not considered, the only decision is when to bring in the challenger.

Since this is a before-tax problem, depreciation and book values are not relevant.

Work in thousands of dollars.

Keep the defender:

There is no drop in salvage value.

The cost for next year is

$$200 + (400)(0.15) = 260$$

Replace the defender:

$$\begin{aligned}\mathrm{EUAC} &= (800)(A/P, 15\%, 10) + 40\\&\quad + (35)(A/G, 15\%, 10)\\&\quad - (150)(A/F, 15\%, 10)\\&= (800)(0.1993) + 40 + (35)(3.3832)\\&\quad - (150)(0.0493)\\&= 159 + 40 + 118 - 7\\&= 310\end{aligned}$$

Since 260 < 310, keep the defender.

Since the defender's economic picture does not change for three years, keep the defender at least that long.

6. Work in thousands of dollars.

Check to see if undiscounted revenues are sufficient to cover costs.

A: $-200 + (14)(30) = 220$

B: $-400 + (8)(10) + (20)(10) + (8)(10) = -40$

C: $-320 + (6)(10) + (6)(10) + (10)(10) = -100$

D: $-300 + (40)(10) + (40)(10) = 500$

Alternatives B and C are not profitable and are eliminated.

Alternative A looks marginal. Check its present worth at 8%.

$$\begin{aligned} P(\mathrm{A}) &= -200 + (14)(P/A, 8\%, 30) \\ &= -200 + (14)(11.2578) \\ &= -42 \ \text{[no good]} \end{aligned}$$

Check alternative D at 8%.

$$\begin{aligned} P(\mathrm{D}) &= -300 + (40)(P/A, 8\%, 10) \\ &\quad + (40)(P/A, 8\%, 30) - (40)(P/A, 8\%, 20) \\ &= -300 + (40)(6.7101 + 11.2578 - 9.8181) \\ &= 26 \ \text{[ok]} \end{aligned}$$

Since present worth at 8% is greater than zero, ROR > 8%.

Choose D.

7. Check to see if ROR > 12% by calculating the present worths at 12%.

A: $$\begin{aligned} P(\mathrm{A}) &= -8000 + (3000)(P/A, 12\%, 4) \\ &= -8000 + (3000)(3.0373) \\ &= 1112 \end{aligned}$$

Since PW > 0, ROR > 12%.

Keep alternative A.

B: $$\begin{aligned} P(\mathrm{B}) &= -10{,}000 + (2500)(P/A, 12\%, 4) \\ &\quad + (500)(P/G, 12\%, 4) \\ &= -10{,}000 + (2500)(3.0373) \\ &\quad + (500)(4.1273) \\ &= -343 \end{aligned}$$

Since PW < 0, ROR < 12%.

Eliminate alternative B.

C: This alternative doesn't pay for itself.

Eliminate alternative C.

C and D: The cash flow diagram is

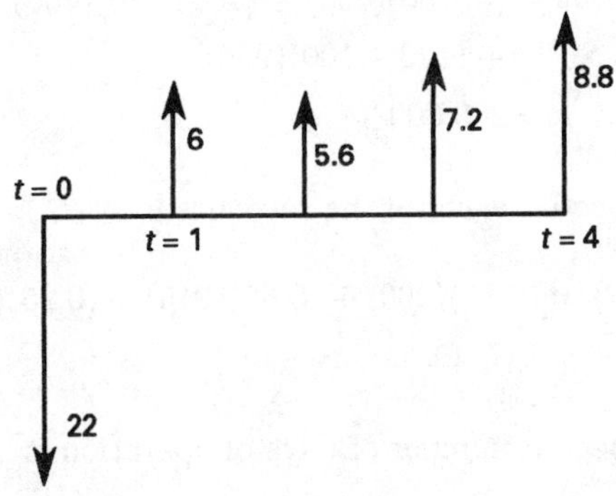

This is the same as

2000 at $t = 1$
4000 every year
1600 gradient

$$\begin{aligned} P(\text{C and D}) &= -22{,}000 + (2000)(P/F, 12\%, 1) \\ &\quad + (4000)(P/A, 12\%, 4) \\ &\quad + (1600)(P/G, 12\%, 4) \\ &= -22{,}000 + (2000)(0.8929) \\ &\quad + (4000)(3.0373) + (1600)(4.1273) \\ &= -1461 \end{aligned}$$

Eliminate alternatives C and D.

Only alternative A survives. If possible, purchase three units of alternative A.

Alternative A

$$(3)(8000) = 24{,}000 < 25{,}000$$

8. Use expected values.

15% of the years:

$$R = \text{revenues} = (5)(80)(500) = 200{,}000$$

60% of the years:

$$\begin{aligned} R &= (5)[(80)(500) + (100 - 80)(450)] \\ &= 245{,}000 \end{aligned}$$

25% of the years:

$$\begin{aligned} R &= (5)[(80)(500) + (100-80)(450) \\ &\quad + (120-100)(350)] \\ &= 280{,}000 \end{aligned}$$

The expected value of the revenues is

$$\begin{aligned} E(R) &= (0.15)(200) + (0.60)(245) + (0.25)(280) \\ &= 247 \quad \text{[thousands]} \end{aligned}$$

The expected number of days of operation is

$$(0.15)(80) + (0.60)(100) + (0.25)(120) = 102$$

The expected annual labor costs are

$$E(C) = (102)(300) = 30.6 \quad \text{[thousands]}$$

The depreciation is

$$D_j = \frac{375-50}{5} = 65$$

The tax credit is

$$(0.10)(375) = 37.5$$

The present worth is

$$\begin{aligned} P &= -375 + (50)(P/F, 25\%, 5) \\ &\quad + (65)(P/A, 25\%, 5)(0.40) \\ &\quad + (37.5)(P/F, 25\%, 1) \\ &\quad - (10)(P/A, 25\%, 5)(1-0.40) \\ &\quad - (30.6)(P/A, 25\%, 5)(1-0.40) \\ &\quad + (247)(P/A, 25\%, 5)(1-0.40) \\ &= -375 + (50)(0.3277) + (65)(2.6893)(0.40) \\ &\quad + (37.5)(0.800) - (10)(2.6893)(0.60) \\ &\quad - (30.6)(2.6893)(0.60) + (247)(2.6893)(0.60) \\ &= -375 + 16.4 + 69.9 + 30 - 16.1 - 49.4 + 398.6 \\ &= 74.4 \quad \text{[thousands]} \end{aligned}$$

(b) **Since PW > 0, the investment is attractive.**

(a) The annualized after-tax income is

$$\begin{aligned} I &= (74.4)(A/P, 25\%, 5) \\ &= (74.4)(0.3718) \\ &= \boxed{27.7 \quad \text{[thousands]}} \end{aligned}$$

(c) The "payback period" usually does not consider interest and other factors. The payback period is

$$\begin{aligned} n &= \frac{375-37.5}{(247-30.6-10)(1-0.40)+(65)(0.40)} \\ &= \boxed{2.25 \text{ years}} \end{aligned}$$

9. (a) The payments adjusted for inflation are

$$\begin{aligned} \text{year 1:} &\quad (1{,}250{,}000)(1.04) = 1{,}300{,}000 \\ \text{year 2:} &\quad (625{,}000)(1.04)^2 = 676{,}000 \\ \text{year 3:} &\quad (625{,}000)(1.04)^3 = 703{,}040 \\ &\quad \text{total } 2{,}679{,}040 \end{aligned}$$

It is assumed that the wholesaler is on the accrual basis for depreciation and income tax purposes.

The depreciation each year is

$$\frac{2{,}679{,}040}{25} = 107{,}162 \quad \text{[say 1.07 hundred thousand]}$$

The combined tax rate is

$$\begin{aligned} t &= s + f - sf \\ &= 0.06 + 0.37 - (0.06)(0.37) \\ &= 0.41 \end{aligned}$$

Work in hundreds of thousands of dollars.

$$\begin{aligned} P &= -(12.5)(1.04)(P/F, 12\%, 1) \\ &\quad - (6.25)(1.04)^2(P/F, 12\%, 2) \\ &\quad - (6.25)(1.04)^3(P/F, 12\%, 3) \\ &\quad + (1.07)(P/A, 12\%, 25)(0.41) \\ &= -(12.5)(1.04)(0.8929) - (6.25)(1.04)^2(0.7972) \\ &\quad - (6.25)(1.04)^3(0.7118) + (1.07)(7.8431)(0.41) \\ &= \boxed{-18.56 \ (1{,}856{,}000)} \end{aligned}$$

(b)

$$\begin{aligned} 18.56 &= A(P/A, 12\%, 25)(1-0.41) \\ &= A(7.8431)(0.59) \\ A &= \boxed{4.01 \ (401{,}000)} \end{aligned}$$

(c) No. The construction company did not correct for its own MARR. It gave the wholesaler a 4% loan, and it should have been at (approximately)

$$\boxed{4\% + 12\% = 16\%}$$

FLUID STATICS AND DYNAMICS

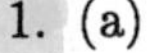
1. (a)

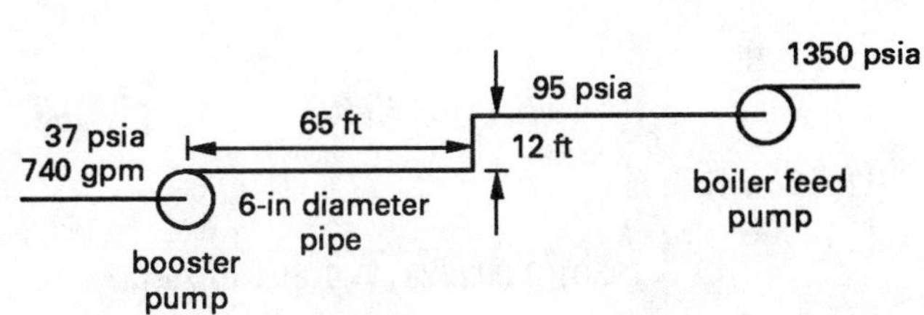

For 6-in pipe,

$$D = 0.5054 \text{ ft}$$
$$A = 0.2006 \text{ ft}^2$$

$$Q = (740)\left(0.002228\,\frac{\frac{\text{ft}^3}{\text{sec}}}{\frac{\text{gal}}{\text{min}}}\right) = 1.649 \text{ ft}^3/\text{sec}$$

$$\text{v} = \frac{Q}{A} = \frac{1.649}{0.2006} = 8.22 \text{ ft/sec}$$

For water at 250°F,

$$\nu = 0.269 \times 10^{-5} \text{ ft}^2/\text{sec}$$
$$\rho = 58.8 \text{ lbm/ft}^3$$

The equivalent length of pipe is

$$L_e = 65 + (4)(5.7) + 3.2 + 63 = 154 \text{ ft}$$

For the steel pipe,

$$\epsilon = 0.0002$$
$$\frac{\epsilon}{D} = \frac{0.0002}{0.5054} \approx 0.0004$$
$$N_{\text{Re}} = \frac{\text{v}D}{\nu} = \frac{(8.22)(0.5054)}{0.269 \times 10^{-5}} = 1.54 \times 10^6$$
$$f = 0.017$$

The friction loss is

$$h_f = \frac{fL\text{v}^2}{2Dg} = \frac{(0.017)(154)(8.22)^2}{(2)(0.5054)(32.2)} = 5.43 \text{ ft}$$

The booster pump discharge pressure is

$$p = 95 + \frac{(12 + 5.43)(58.8)}{144} = \boxed{102.1 \text{ psia}}$$

(b) $P = \dot{m}\Delta h$

$$= \frac{(1.649)(58.8)\left(\frac{1350 - 95}{58.8}\right)(144)}{(0.62)(550)} = \boxed{874 \text{ hp}}$$

2. From affinity laws,

(a)

$$Q \propto n$$
$$H \propto n^2$$
$$\text{NPSHR} \propto n^2$$

The ratio of speeds is

$$\frac{n_{\text{new}}}{n_{\text{old}}} = \frac{2000}{1800} = 1.11$$

$(1.11)(Q)$	$(1.11)^2(H)$	$(1.11)^2(\text{NPSHR})$
611	61.6	9.9
666	58.5	11.7
722	55.4	13.7
777	51.9	16.0
833	48.2	18.5
888	44.4	21.2

These values can be graphed if necessary.

(b) For water at 90°F,

$$\rho = 62.11 \text{ lbm/ft}^3$$
$$h_{\text{vp}} = 1.61 \text{ ft}$$

At 5000 ft,

$$p_a = 12.225 \text{ psia}$$
$$h_a = \frac{(12.225)(144)}{62.11} = 28.34 \text{ ft}$$

The net positive suction head is

$$\text{NPSHA} = 28.34 + (-7) - 1.61 - 10 = 9.73 \text{ ft}$$

Since $9.73 < 11.1$,

Operation is not ok.

Notice that the discharge head is not considered.

3. (a)

$$h_1 = 188.13 \text{ BTU/lbm}$$
$$h_2 = 157.95 \text{ BTU/lbm}$$
$$\Delta h = 188.13 - 157.95 = 30.18 \text{ BTU/lbm}$$
$$\dot{m} = \frac{400{,}000}{(30.18)(3600)} = 3.68 \text{ lbm/sec}$$
$$\rho_1 = \frac{1}{0.01677} = 59.63 \text{ lbm/ft}^3$$
$$\nu_1 = 0.312 \times 10^{-5} \text{ ft}^2/\text{sec}$$

ν_1 is interpolated for 220°F water. An argument for using $\left(\frac{1}{2}\right)(220+190) = 205$°F water could also be made.

For the pipe,

$$A = 0.03325 \text{ ft}^2$$
$$D = 0.2058 \text{ ft}$$
$$\text{v} = \frac{Q}{A} = \frac{\frac{3.68}{59.63}}{0.03325} = 1.856 \text{ ft/sec}$$

The equivalent length is

$$L_e = 225 + (8)(9.3) + 22 + (2)(1.7)$$
$$= 324.8 \text{ ft}$$
$$\epsilon = 0.0002$$
$$\frac{\epsilon}{D} = \frac{0.0002}{0.2058} \approx 0.001$$
$$N_{\text{Re}} = \frac{\text{v}D}{\nu} = \frac{(1.856)(0.2058)}{0.312 \times 10^{-5}}$$
$$= 1.2 \times 10^5$$
$$f = 0.022$$
$$h_f = 12 + \frac{(0.022)(324.8)(1.856)^2}{(2)(0.2058)(32.2)}$$
$$= 13.9 \text{ ft}$$

The power is

$$P = \dot{m}h = \frac{(3.68)(13.9)}{550}$$
$$= \boxed{0.093 \text{ hp}}$$

(b) The motor power required is

$$\frac{0.093}{(0.55)(0.75)} = 0.23 \quad [\text{say } \tfrac{1}{4} \text{ hp}]$$

(In practice, a $\frac{1}{3}$-hp motor would probably be chosen.)

Converting maximum power to watts,

$$P = \left(\tfrac{1}{4} \text{ hp}\right)\left(745.7 \frac{\text{W}}{\text{hp}}\right) = 186.4 \text{ W}$$

$$\boxed{\text{Say 190 W.}}$$

4. For water at 180°F,

$$\rho = 60.58 \text{ lbm/ft}^3$$
$$h_{\text{vp}} = 17.33 \text{ ft}$$

At 145°F,

$$h_{\text{vp}} \approx 7.63 \text{ ft}$$

The flow rate is

$$Q = (850)(0.002228) = 1.894 \text{ ft}^3/\text{sec}$$
$$D_i = \frac{6}{12} = 0.5 \text{ ft}$$
$$A_i = \left(\frac{\pi}{4}\right)(0.5)^2 = 0.1963$$
$$\text{v} = \frac{Q}{A} = \frac{1.894}{0.1963} = 9.65 \text{ ft/sec}$$

(a) Assume $c_p = 1.0$.

$$q = mc_p\Delta T = (1.894)(60.58)(180 - 145)(3600)$$
$$= \boxed{1.446 \times 10^7 \text{ BTU/hr}}$$

(b) The velocity head is

$$h_\text{v} = \frac{\text{v}^2}{2g} = \frac{(9.65)^2}{(2)(32.2)} = 1.45 \text{ ft}$$

The pressure head is

$$h_p = \frac{p}{\gamma} = \frac{(14.1)(144)}{60.58} = 33.52 \text{ ft}$$

The NPSHA is

$$\text{NPSHA} = 33.52 + 1.45 - 17.33$$
$$= 17.64 \text{ ft}$$

Since $17.64 < 21$,

$$\boxed{\text{The pump will cavitate.}}$$

(c) If the vapor pressure head is reduced,

$$\text{NPSHA} = 33.52 + 1.45 - 7.63$$
$$= 27.34 \text{ ft}$$

Since $27.34 > 21$,

$$\boxed{\text{The pump will not cavitate.}}$$

5. Since the diameter is not known to be inside or outside, no attempt is made to distinguish between internal and external volumes.

The volume displaced by the pipeline is

$$\left(\frac{\pi}{4}\right)\left(\frac{30}{12}\right)^2(325) = 1595.3 \text{ ft}^3$$

The buoyant force on the pipeline is

$$(1595.3)(1.36)(62.4) = 135{,}384 \text{ lbf}$$

The buoyant force on the concrete anchors is

$$\left(\frac{12{,}000}{140}\right)(1.36)(62.4) = 7274 \text{ lbf each}$$

The density of steel is 489 lbm/ft^3. The weight of the pipeline is

$$\pi\left(\frac{30}{12}\right)\left(\frac{0.5}{12}\right)(325)(489) = 52{,}008$$

The weight of the gas is calculated as follows.

$$m = V\rho$$
$$V = 1595.3$$
$$\rho = \frac{p}{RT} = \frac{(900)(144)}{\left(\frac{1545}{18.9}\right)(460+73)} = 2.974 \text{ lbm/ft}^3$$
$$\text{weight} \approx m = (1595.3)(2.974) = 4744$$

The factor of safety is defined as

$$\text{FS} = \frac{\text{total weight}}{\text{total buoyant force}}$$

Let n be the number of concrete anchors. Choose FS = 1.2 (*Gas Engineer's Handbook*).

$$1.2 = \frac{52{,}008 + 4744 + (n)(12{,}000)}{135{,}384 + (n)(7274)}$$
$$n = 32.3 \text{ [say 32.5]}$$
$$\text{spacing} = \frac{325}{32.5} = \boxed{10 \text{ ft}}$$

Other points:

- A higher factor of safety actually can be achieved with fewer anchors. For example, FS = 3.0 when $n = 25$. Optimization may be required.
- The minimum factor of safety must be satisfied when the pipeline is empty. It would be more conservative to omit the gas weight from the calculation. However, 4744 does not significantly affect the answer and is included for completeness.

6. (a) Solving for v_o,

$$\text{v}_o = \frac{x}{\sqrt{\frac{2y}{g}}} = \frac{8.5}{\sqrt{\frac{(2)\left(\frac{26}{12}\right)}{32.2}}}$$
$$= 23.17 \text{ ft/sec}$$
$$A_o = \left(\frac{\pi}{4}\right)\left(\frac{3.5}{12}\right)^2 = 0.0668 \text{ ft}^2$$
$$Q = \text{v}_o A_o = \frac{\left(23.17\ \frac{\text{ft}}{\text{sec}}\right)(0.0668 \text{ ft}^2)}{0.002228\ \frac{\text{ft}^3\text{-min}}{\text{sec-gal}}}$$
$$= \boxed{694.7 \text{ gal/min}}$$

(b) The head in the main is calculated as follows.

$$x = 2C_v\sqrt{hy}$$
$$8.5 = (2)(0.82)\sqrt{h}\sqrt{\frac{26}{12}}$$
$$h = 12.4 \text{ ft}$$

The pressure is

$$p = \gamma h = \frac{(62.4)(12.4)}{144} = \boxed{5.37 \text{ psig}}$$

(Seems low. 60 psig is normal.)

7. For the pipe,

$$D = 0.835 \text{ ft}$$
$$A = 0.5476 \text{ ft}^2$$

For 90°F water,

$$\rho = 62.11 \text{ lbm/ft}^3$$
$$\nu = 0.826 \times 10^{-5}$$

The equivalent length is

$$L_e = 380 + (10)(8) + (4)(14) + (4)(9) + (1)(3.2) + (1)(310) + (2)(5.2) + (1)(30)$$
$$= 905.6 \text{ ft}$$

The flow velocity is

$$\text{v} = \frac{Q}{A} = \frac{(1900)(0.002228)}{0.5476} = 7.73 \text{ ft/sec}$$
$$N_{\text{Re}} = \frac{\text{v}D}{\nu} = \frac{(7.73)(0.835)}{0.826 \times 10^{-5}} = 7.8 \times 10^5$$

For steel pipe,

$$\epsilon = 0.0002$$
$$\frac{\epsilon}{D} = \frac{0.0002}{0.835} = 0.00024$$

From the Moody diagram, $f = 0.016$.

$$h_f = \frac{(0.016)(905.6)(7.73)^2}{(2)(0.835)(32.2)} = 16.1 \text{ ft}$$

From the Bernoulli equation,

$$\frac{(155)(144)}{62.11} + h_v + 0 = \frac{(p_2)(144)}{62.11} + h_v - 45 + 16.1$$
$$p_2 = 167.5 \text{ psig} \quad [\text{higher than } p_1]$$

(a) The static pressure head gain is

$$\frac{(167.5 - 155)(144)}{62.11} = \boxed{29.0 \text{ ft}}$$

(b) Although there is a static pressure gain, this comes from a drop in elevation. There is an energy loss of 16 ft.

8. The friction loss is

$$h_f = \frac{fL_e\text{v}^2}{2Dg_c} = \frac{fL_e\left(\frac{\dot{m}}{\rho A}\right)^2}{2Dg_c}$$
$$= \frac{fL_e\left[\frac{\dot{m}}{\rho\left(\frac{\pi}{4}\right)D^2}\right]^2}{2Dg_c} = \frac{0.811fL_e\dot{m}^2}{\rho^2D^5g_c}$$

Solving for D,

$$D = \sqrt[5]{\frac{0.811fL_e\dot{m}^2}{g_ch_f\rho^2}}$$

This problem is one of finding f (which varies with N_{Re}).

Equivalent length:

Get a preliminary size by assuming a reasonable velocity (say 5 ft/sec).

$$A = \frac{\dot{m}}{\rho\text{v}} = \frac{250{,}000}{(3600)(1.45)(62.4)(5)} = 0.154 \text{ ft}^2$$
$$D = \sqrt{\frac{4A}{\pi}} = \sqrt{\frac{(4)(0.154)}{\pi}}$$
$$= 0.44 \text{ ft} \quad [5.3 \text{ in—say } 6 \text{ in}]$$

The equivalent length is

$$L_e = 1500 + (2)(3.2) + (2)(5.7)$$
$$= 1517.8 \quad [\text{say } 1518 \text{ ft}]$$

Friction loss:

The allowable head loss is

$$h_f = \frac{p}{\gamma} = \frac{(23 - 18)(144)}{(1.45)(62.4)} = 7.96 \text{ ft of solution}$$

At 5 ft/sec and 6-in pipe,

$$N_{\text{Re}} = \frac{D\text{v}}{\nu} = \frac{\left(\frac{6}{12}\right)(5)}{1.2 \times 10^{-4}} = 2.1 \times 10^4$$

For steel,

$$\epsilon = 0.0002$$
$$\frac{\epsilon}{D} = \frac{0.0002}{\frac{6}{12}} = 0.0004$$

From the Moody friction factor chart,

$$f \approx 0.025$$

Solving for the diameter,

$$D = \sqrt[5]{\frac{(250{,}000)^2(0.811)(0.025)(1518)}{(3600)^2[(1.45)(62.4)]^2(32.2)(7.96)}}$$
$$= 0.589 \text{ ft}$$

$$\boxed{\text{Use 8-in pipe } (D = 0.6651 \text{ ft}).}$$

(b) Solve for h_f using 8-in pipe.

$$D = 0.6651 \text{ ft}$$
$$A = 0.3474 \text{ ft}^2$$
$$L_e = 1518 \text{ ft} \quad [\text{essentially unchanged}]$$
$$\text{v} = \frac{\dot{m}}{\rho A} = \frac{250{,}000}{(3600)(1.45)(62.4)(0.3474)}$$
$$= 2.21 \text{ ft/sec}$$
$$\epsilon = 0.0002$$
$$\frac{\epsilon}{D} = \frac{0.0002}{0.6651} = 0.0003$$
$$N_{\text{Re}} = \frac{\text{v}D}{\nu} = \frac{(2.21)(0.6651)}{1.2 \times 10^{-4}} = 1.2 \times 10^4$$
$$f = 0.030$$
$$h_f = \frac{fL\text{v}^2}{2Dg} = \frac{(0.03)(1518)(2.21)^2}{(2)(0.6651)(32.2)}$$
$$= 5.19 \text{ ft}$$
$$p = \gamma h = \frac{(1.45)(62.4)(5.19)}{144}$$
$$= \boxed{3.26 \text{ psi}}$$

9. $T_{°F} = 32 + \left(\frac{9}{5}\right)(5) = 41°\text{F}$ [say 40 °F]

At 40°F,

$$\nu = 14.6 \times 10^{-5}\ \text{ft}^2/\text{sec}$$

The dimensions of each passageway are

$$\begin{aligned}\text{height} &= 1.25 - (2)(0.0625) = 1.125\ \text{in}\\ \text{width} &= 12.00 - (73+1)(0.0625) = 7.375\ \text{in}\\ \frac{7.375}{73} &= 0.101\ \text{in}\end{aligned}$$

The area in flow is

$$\begin{aligned}A &= \frac{(2\ \text{layers})(1.125)(7.375)}{144}\\ &= 0.1152\ \text{ft}^2\end{aligned}$$

The velocity is

$$\text{v} = \frac{Q}{A} = \frac{\frac{20}{60}}{0.1152} = 2.9\ \text{ft/sec}$$

The equivalent diameter of the passageway is

$$\begin{aligned}D_e &= \left(\frac{1}{12}\right)\left[\frac{(2)(1.125)(0.101)}{1.125+0.101}\right]\\ &= 0.01545\ \text{ft}\end{aligned}$$

The Reynolds number is

$$N_{\text{Re}} = \frac{\text{v}D}{\nu} = \frac{(0.01545)(2.9)}{14.6\times 10^{-5}} = 307\ \text{[laminar]}$$

Material and rougness are irrelevant.

$$f = \frac{64}{N_{\text{Re}}} = \frac{64}{307} = 0.208$$

$$\begin{aligned}h_f &= \frac{fL\text{v}^2}{2Dg} = \frac{(0.208)\left(\frac{12}{12}\right)(2.9)^2}{(2)(0.01545)(32.2)}\\ &= 1.758\ \text{ft of air}\end{aligned}$$

This is a parallel drop across all of the passageways.

Add in the minor losses.

$$\begin{aligned}h_f &= 1.758 + (0.52+0.9)\left[\frac{(2.9)^2}{(2)(32.2)}\right]\\ &= 1.94\ \text{ft of air}\end{aligned}$$

Convert to inches of water. (There has been no attempt to use exact densities.) $\rho_{\text{air}} \approx 0.075\ \text{lbm/ft}^3$.

$$\begin{aligned}h_f &= (1.94)\left(12\ \frac{\text{in}}{\text{ft}}\right)\left(\frac{0.075\ \frac{\text{lbm}}{\text{ft}^3}}{62.4\ \frac{\text{lbm}}{\text{ft}^3}}\right)\\ &= \boxed{0.028\ \text{in w.g.}}\end{aligned}$$

THERMODYNAMICS

1. It is possible to determine the fraction of each gas from the properties (R, c_p, etc.), but it is not necessary.

The isentropic efficiency affects the ideal power.

$$P_{\text{ideal}} = \frac{5.2}{0.70} = 7.429 \text{ hp}$$

The mechanical efficiency increases the mass flow rate.

$$\dot{m}_{\text{actual}} = \frac{\dot{m}_{\text{ideal}}}{\eta_{\text{mechanical}}}$$

$$\begin{aligned}\dot{m}_{\text{ideal}} &= \dot{m}_{\text{actual}}\eta_{\text{mechanical}} \\ &= (340)(0.85) \\ &= 289 \text{ lbm/hr}\end{aligned}$$

The expression for power is

$$P = \frac{\dot{m}\Delta h J}{550} \quad \text{[in horsepower]}$$

Solving for the enthalpy change,

$$\begin{aligned}\Delta h_{\text{ideal}} &= \frac{\left(550 \frac{\text{ft-lbf}}{\text{sec-hp}}\right)(7.429 \text{ hp})\left(3600 \frac{\text{sec}}{\text{hr}}\right)}{\left(289 \frac{\text{lbm}}{\text{hr}}\right)\left(778 \frac{\text{ft-lbf}}{\text{BTU}}\right)} \\ &= 65.42 \text{ BTU/lbm} \quad \text{[isentropic]}\end{aligned}$$

(a) Assume constant gas properties and adiabatic operation.

$$W = \Delta h = c_p T_1 \left[1 - \left(\frac{p_2}{p_1}\right)^{\frac{k-1}{k}}\right]$$

$$65.42 = (0.218)(240 + 460)\left[1 - \left(\frac{16}{p_1}\right)^{\frac{1.355-1}{1.355}}\right]$$

$$0.429 = 1 - \left(\frac{16}{p_1}\right)^{0.262}$$

$$p_1 = \boxed{135.6 \text{ psi}}$$

Notice that the isentropic expansion equation could only be used with the isentropic power.

Also, p_1 and p_2 do not depend on the process being isentropic.

(b) $\eta_{\text{overall}} = (0.70)(0.85) = \boxed{0.595\ (59.5\%)}$

(c) s_1 and s_2 are not known, so Δs must be found directly.

Note that $\Delta T_{°\text{F}} = \Delta_{°\text{R}}$,

$$\Delta T_{\text{actual}} = \frac{\Delta h_{\text{actual}}}{c_p}$$

$$\begin{aligned}T_2 &= T_1 - \frac{\Delta h_{\text{actual}}}{c_p} \\ &= 240 - \frac{(65.42)(0.70)}{0.218} = 30°\text{F}\end{aligned}$$

Notice that T_2 must be calculated from h_{actual} since temperature depends on whether the process is isentropic.

$$\begin{aligned}\Delta s &= c_p \ln\left(\frac{T_2}{T_1}\right) - R\ln\left(\frac{p_2}{p_1}\right) \\ &= 0.218 \ln\left(\frac{30 + 460}{240 + 460}\right) - 0.057 \ln\left(\frac{16}{135.6}\right) \\ &= \boxed{0.044 \text{ BTU/lbm-°R}}\end{aligned}$$

(d)

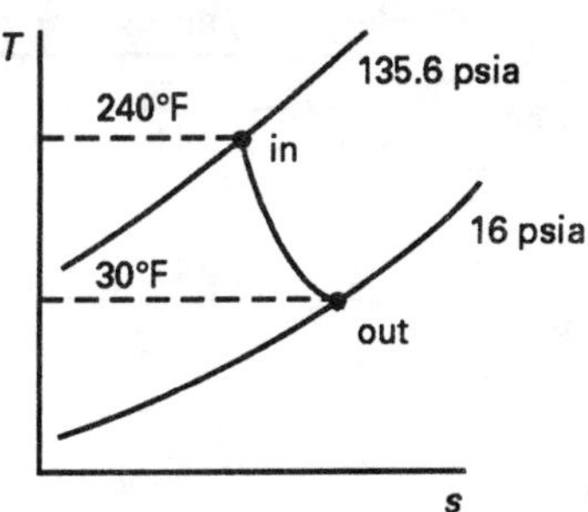

2.

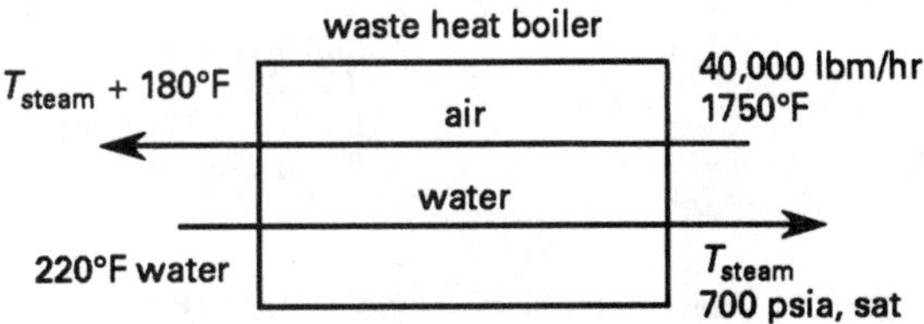

(a) T_{sat} for 700 psia is 503.10°F.

$$\begin{aligned}h_{\text{steam,out}} &= 1201.2 \text{ BTU/lbm} \\ h_{\text{water,in}} &= 188.13 \text{ BTU/lbm} \\ T_{\text{air,out}} &= 503 + 180 = 683°\text{F}\end{aligned}$$

At this temperature, assuming constant c_p probably would be inappropriate.

$$q_{\text{air}} = \dot{m}\Delta h$$

From the air table,

$$\begin{aligned} T_{\text{air,in}} &= 1750 + 460 = 2210°\text{R} \\ h_{\text{air,in}} &= 563.4 \text{ BTU/lbm} \\ T_{\text{air,out}} &= 683 + 460 = 1143°\text{R} \\ h_{\text{air,out}} &= 276.8 \text{ BTU/lbm} \\ q_{\text{air}} &= \dot{m}_{\text{air}}\Delta h = (40{,}000)(563.4 - 276.8) \\ &= 1.146 \times 10^7 \text{ BTU/hr} \end{aligned}$$

Assuming no friction and an adiabatic process,

$$\begin{aligned} q_{\text{air}} &= q_{\text{steam}} \\ 1.146 \times 10^7 &= \dot{m}_{\text{steam}}(1201.2 - 188.13) \\ \dot{m}_{\text{steam}} &= \boxed{11{,}312 \text{ lbm/hr}} \end{aligned}$$

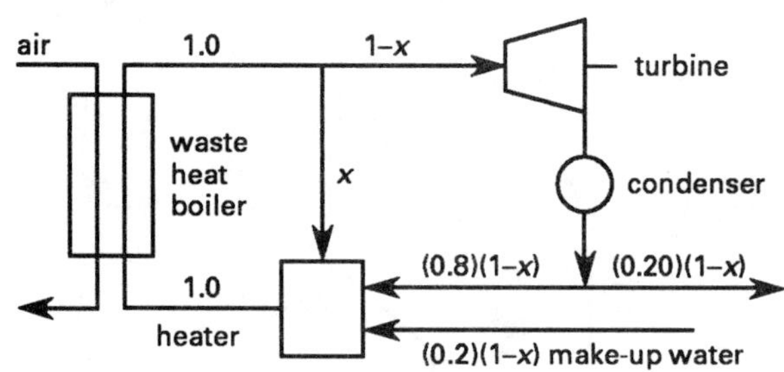

(b)

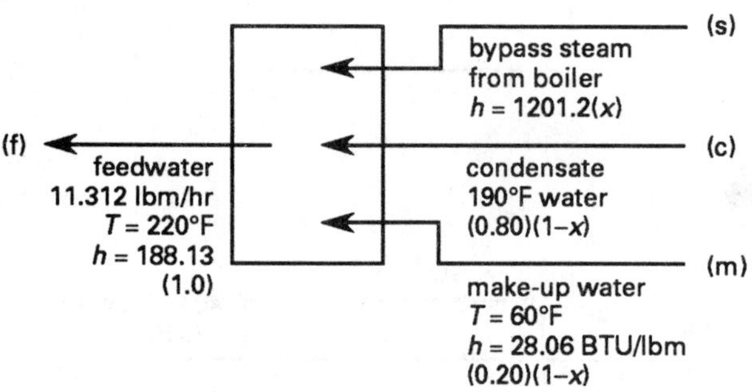

$$h_{\text{condensate}} = 157.95 \text{ BTU/lbm}$$

Assuming adiabatic operation,

$$\begin{aligned} E_f &= E_s + E_c + E_m \\ m_f h_f &= m_s h_s + m_c h_c + m_m h_m \end{aligned}$$

Expressing this per pound of steam ($m_s = 1$),

$$\begin{aligned} h_f &= x h_s + (0.80)(1 - x)h_c \\ &\quad + (0.20)(1 - x)h_m \\ h_f - 0.8h_c - 0.2h_m &= x h_s - 0.8x h_c - 0.2x h_m \end{aligned}$$

$$\begin{aligned} x &= \frac{h_f - 0.2h_m - 0.8h_c}{h_s - 0.2h_m - 0.8h_c} \\ &= \frac{188.13 - (0.20)(28.06) - (0.8)(157.95)}{1201.2 - (0.20)(28.06) - (0.8)(157.95)} \\ &= 0.0525 \end{aligned}$$

The mass of steam is

$$m_s = x m_f = (0.0525)(11{,}312) = \boxed{594 \text{ lbm/hr}}$$

(c)

$$\begin{aligned} m_m &= (0.20)(1 - x)m_f \\ &= (0.20)(1 - 0.0525)(11{,}312) \\ &= \boxed{2144 \text{ lbm/hr}} \end{aligned}$$

3. (a) Gather the enthalpies.

$$\begin{aligned} h_1 &= 250.09 \text{ BTU/lbm} \\ h_2 &= 127.89 \text{ BTU/lbm} \\ h_3 &= 181.11 + (0.80)(969.7) = 956.87 \text{ BTU/lbm} \\ h_4 &= 180.07 \text{ BTU/lbm} \end{aligned}$$

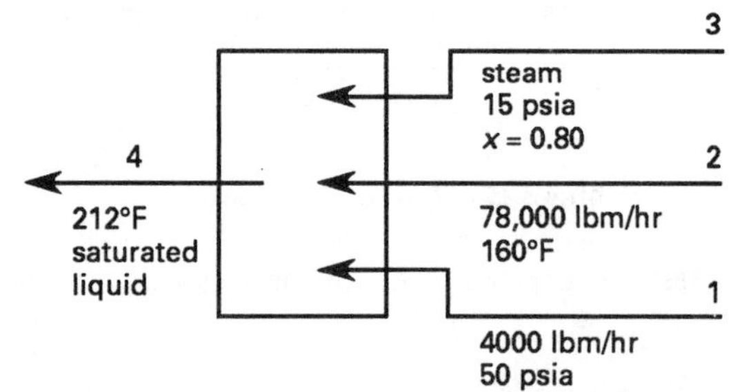

Mass balance:

$$\begin{aligned} m_1 + m_2 + m_3 &= m_4 \\ 4000 + 78{,}000 + m_3 &= m_4 \\ 82{,}000 + m_3 &= m_4 \end{aligned}$$

Energy balance:

$$m_1 h_1 + m_2 h_2 + m_3 h_3 = m_4 h_4 + 225{,}000$$

But,

$$\begin{aligned} m_4 &= 82{,}000 + m_3 \\ m_1 h_1 + m_2 h_2 + m_3 h_3 &= (82{,}000 + m_3)h_4 + 225{,}000 \\ m_1 h_1 + m_2 h_2 + m_3(h_3 - h_4) &= 82{,}000 h_4 + 225{,}000 \end{aligned}$$

$$\begin{aligned} m_3 &= \frac{\begin{array}{c}(82{,}000)(180.07) + 225{,}000 \\ -(4000)(250.09) - (78{,}000)(127.89)\end{array}}{956.87 - 180.07} \\ &= \boxed{5168.6 \text{ lbm/hr}} \end{aligned}$$

(b)
$$\begin{aligned} m_4 &= m_1 + m_2 + m_3 \\ &= 4000 + 78{,}000 + 5168.6 \\ &= 87{,}168.6 \text{ lbm/hr} \end{aligned}$$

At saturated 212°F,

$$v_F = 0.01672 \text{ ft}^3/\text{lbm}$$
$$\text{volume} = (87{,}168.6)(0.01672) = 1457.5 \text{ ft}^3/\text{hr}$$
$$= \frac{\left(1457.5 \ \frac{\text{ft}^3}{\text{hr}}\right)(7.4805)}{60 \ \frac{\text{min}}{\text{hr}}} = \boxed{181.7 \text{ gal/min}}$$

(c) Looking at the equation for m_3 and m_4, it is not necessary to recalculate everything.

$$\text{decrease in } m_4 = \text{decrease in } m_3$$
$$\begin{aligned} \Delta m_4 &= \frac{\Delta q}{\Delta h} = \frac{225{,}000 - 110{,}000}{956.87 - 180.07} \\ &= 148.0 \text{ lbm/hr} \end{aligned}$$
$$\begin{aligned} \text{flow rate} &= 181.7 - \frac{(148)(0.01672)(7.4805)}{60} \\ &= \boxed{181.4 \text{ gal/min}} \end{aligned}$$

4.

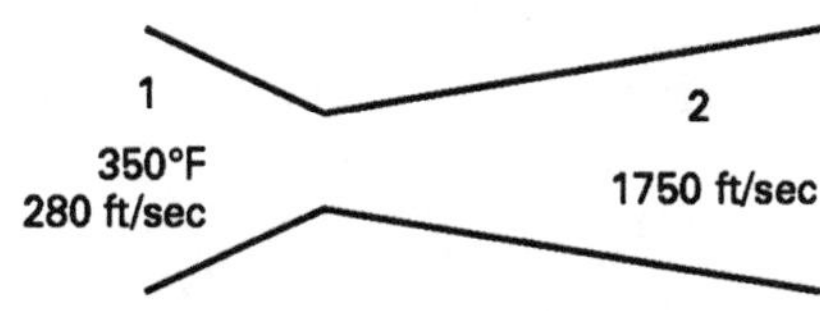

(a) The velocity increase comes from a drop in enthalpy.

$$h_1 = 1192.3 \text{ BTU/lbm}$$

Since this is an observed velocity,

$$\begin{aligned} h_{2,\text{actual}} &= h_1 + \frac{\text{v}_1^2 - \text{v}_2^2}{2gJ} \\ &= 1192.3 + \frac{(280)^2 - (1750)^2}{(2)(32.2)(778)} \\ &= \boxed{1132.7 \text{ BTU/lbm}} \end{aligned}$$

(c) If the expansion had been perfect, a larger enthalpy drop would have been realized.

$$\begin{aligned} h_{2,\text{ideal}} &= 1192.3 - \frac{1192.3 - 1132.7}{0.90} \\ &= 1126.1 \text{ BTU/lbm} \end{aligned}$$

At point 1,

$$T_1 = 350°\text{F}$$
$$p_1 = 134.63 \text{ psia}$$
$$h_1 = 1192.3 \text{ BTU/lbm}$$

From the Mollier diagram, dropping straight down from point 1 to $h_2 = 1126.1$, the pressure is

$$p_2 = \boxed{58 \text{ psia}}$$

(b) At saturation and 58 psia,

$$h_F = 259.82 \text{ BTU/lbm}$$
$$h_{FG} = 917.1 \text{ BTU/lbm}$$

Since $h = h_F + xh_{FG}$,

$$x = \frac{1132.7 - 259.82}{917.1} = \boxed{0.952}$$

5. (a) According to Dalton's law of partial pressure, the total pressure is the sum of partial pressures.

$$p_\text{room} = p_\text{air} + p_\text{Halon}$$
$$p_\text{air} = (14.7)(144) = 2116.8 \text{ psfa}$$
$$m_\text{Halon} = \frac{pV}{RT} = \frac{(2116.8)(2300)}{RT}$$
$$\begin{aligned} p_\text{Halon} &= \frac{mRT}{V} = \frac{(2116.8)(2300)(RT)}{(30{,}000)(RT)} \\ &= 162.3 \text{ psfa} \end{aligned}$$
$$\begin{aligned} p_\text{room} &= 2116.8 + 162.3 = \boxed{2279.1 \text{ psfa}} \\ &= \boxed{15.83 \text{ psia}} \end{aligned}$$

(b) Overpressure is found from $p = \gamma h$.

$$\gamma_\text{water} = 0.0361 \text{ lbf/in}^3$$
$$\text{overpressure} = \frac{15.83 - 14.7}{0.0361} = \boxed{31.3 \text{ inches of water}}$$

Disregarding any changes in gas volumes due to combustion and disregarding any decomposition of Halon, overpressure exceeds the 1 in w.g. spec.

(c) Venting is probably by use of spring-loaded or vacuum breaker types of pressure release valves, although electronically controlled valves or rupture diagrams could be used.

Due to the possibility of radioactive particles in the smoke, venting through a scrubber or washer is preferred over direct venting to the atmosphere.

6.

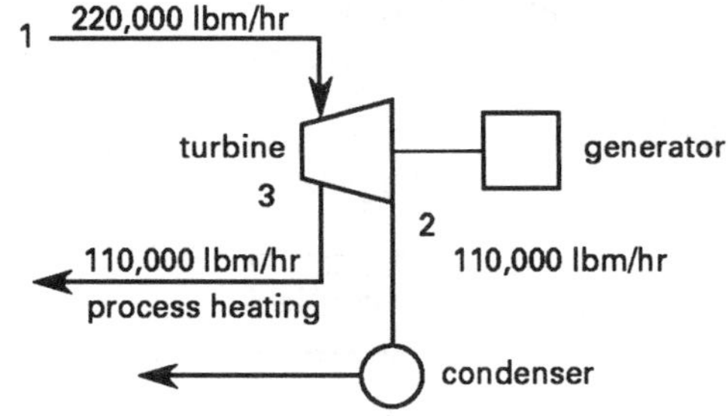

At 1:

$$h_1 = 1444.5 \text{ BTU/lbm}$$

At 2: From the Mollier diagram at $p = 2$ in Hg,

$$h_2 = 893 \text{ BTU/lbm} \quad \text{[if isentropic]}$$

This is inside the vapor dome.

$$\begin{aligned} h_2' &= 1444.5 - (0.75)(1444.5 - 893) \\ &= 1030.9 \text{ BTU/lbm} \quad \text{[say 1031 BTU/lbm]} \end{aligned}$$

(b) At 3:

From the Mollier diagram at $p = 200$ psia,

$$\begin{aligned} h_3 &= 1246 \text{ BTU/lbm} \quad \text{[if isentropic]} \\ h_3' &= 1444.5 - (0.75)(1444.5 - 1246) \\ &= 1295.6 \text{ BTU/lbm} \end{aligned}$$

Say 1296 BTU/lbm.

This is outside the vapor dome.

(a) From the Mollier diagram for $p = 200$ psia and $h = 1296$ BTU/lbm,

$$\boxed{T \approx 550°\text{F}}$$

(c) The power is

$$P = \frac{0.98[(220{,}000)(1444.5 - 1296) + (110{,}000)(1296 - 1031)]}{\left(3412.9 \ \frac{\frac{\text{BTU}}{\text{hr}}}{\text{kW}}\right)\left(1000 \ \frac{\text{kW}}{\text{MW}}\right)}$$

$$= \boxed{17.75 \text{ MW}}$$

7.

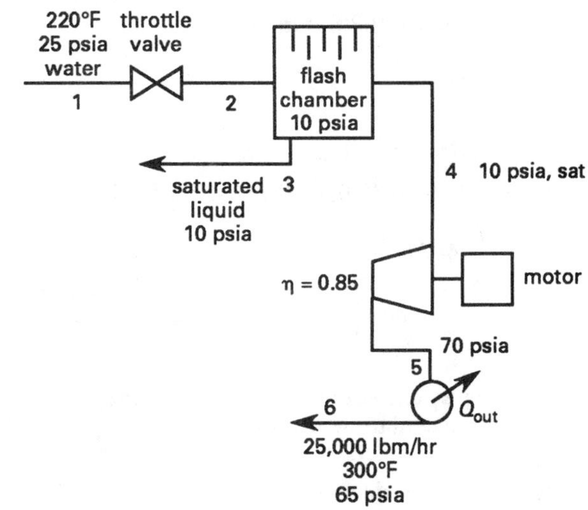

$$h_1 = 188.13 \text{ BTU/lbm}$$

Assume throttling occurs such that $\Delta h = 0$.

$$\begin{aligned} h_2 &= 188.13 \text{ BTU/lbm} \\ p_2 &= 10 \text{ psia} \end{aligned}$$

At 10 psia,

$$\begin{aligned} h_3 &= h_F = 161.17 \text{ BTU/lbm} \\ h_4 &= 1143.3 \text{ BTU/lbm} \end{aligned}$$

Assuming 1 pound entering the flash chamber, the split is

$$m_4 = m_2 - m_3 = 1 - m_3$$

The energy balance is

$$\begin{aligned} m_2 h_2 &= m_3 h_3 + m_4 h_4 \\ (1)(188.13) &= m_3(161.17) + (1 - m_3)(1143.3) \\ m_3 &= 0.973 \quad [97.3\%] \end{aligned}$$

At point 5, from the Mollier diagram at 70 psia,

$$\begin{aligned} h_5 &= 1320 \text{ BTU/lbm} \quad \text{[if isentropic]} \\ h_5' &= 1143.3 + \frac{1320 - 1143.3}{0.85} \\ &= 1351.2 \text{ BTU/lbm} \\ h_6 &= 269.59 \text{ BTU/lbm} \end{aligned}$$

$$\text{(a)} \quad W_{\text{compression}} = \left(25{,}000 \ \frac{\text{lbm}}{\text{hr}}\right) \times \left(1351.2 \ \frac{\text{BTU}}{\text{lbm}} - 1143.3 \ \frac{\text{BTU}}{\text{lbm}}\right) = 5.2 \times 10^6 \text{ BTU/hr}$$

Using a 100% motor efficiency,

$$W_{\text{motor}} = \frac{5.2 \times 10^6 \ \frac{\text{BTU}}{\text{hr}}}{3412.9 \ \frac{\text{BTU}}{\text{hr-kW}}} = \boxed{1524 \text{ kW}}$$

(b)

$$\begin{aligned}
m_1 &= \frac{25{,}000}{1 - 0.973} \\
&= 9.26 \times 10^5 \text{ lbm/hr} \\
Q_{\text{in},1} &= m_1 h_1 \\
&= (9.26 \times 10^5)(188.13) \\
&= 1.74 \times 10^8 \text{ BTU/hr} \\
m_3 &= (0.973)(9.26 \times 10^5) \\
&= 9.01 \times 10^5 \text{ lbm/hr} \\
Q_{\text{out},3} &= m_3 h_3 \\
&= (9.01 \times 10^5)(161.17) \\
&= 1.45 \times 10^8 \text{ BTU/hr} \\
W_{\text{in},4\text{-}5} &= 5.2 \times 10^6 \text{ BTU/hr}
\end{aligned}$$

$$\begin{aligned}
Q_{\text{out},5\text{-}6} &= m_5(h_5 - h_6) \\
&= (25{,}000)(1351.2 - 269.59) \\
&= 2.7 \times 10^7 \text{ BTU/hr} \\
W_{\text{out},6} &= (25{,}000)(269.59) \\
&= 6.74 \times 10^6 \text{ BTU/hr} \\
\eta_{\text{th}} &= \frac{Q_{\text{in}} - Q_{\text{out}}}{Q_{\text{in}}} \\
&= \frac{1.74 \times 10^8 - 1.45 \times 10^8 - 2.7 \times 10^7}{1.74 \times 10^8} \\
&= \boxed{0.01} \quad \text{[accurate enough with rounding errors]}
\end{aligned}$$

The same answer is obtained from

$$\eta_{\text{th}} = \frac{W_{\text{out}} - W_{\text{in}}}{Q_{\text{in}}}$$

Economics, complexity, maintenance, and value of discarded energy are also significant criteria.

POWER CYCLES

1. Assume ideal gas relationships hold.

$$\begin{aligned} q_{\text{to steam}} &= \eta \dot{m}\Delta h = \eta \dot{m} c_p \Delta T \\ &= (0.90)(9)(0.239)(1040 - 650) \\ &= 755 \text{ BTU/sec} \end{aligned}$$

At point 2, from the superheat table or Mollier diagram,

$$h_2 = 1508 \text{ BTU/lbm}$$

At point 3, from the Mollier diagram,

$$\begin{aligned} h_3 &= 967 \text{ BTU/lbm} \quad \text{[if isentropic]} \\ h_3' &= 1508 - (0.80)(1508 - 967) \\ &= 1075 \text{ BTU/lbm} \end{aligned}$$

At point 4, from the saturated steam table at 2 psia,

$$\begin{aligned} h_4 &= 94 \text{ BTU/lbm} \\ \upsilon_{F,4} &= 0.01623 \text{ ft}^3/\text{lbm} \end{aligned}$$

At point 1,

$$\begin{aligned} h_1' &= 94 + \frac{(0.01623)(900 - 2)(144)}{(778)(0.60)} \\ &= 98.5 \text{ BTU/lbm} \end{aligned}$$

(b) The steam flow rate is

$$\begin{aligned} \dot{m}_{\text{steam}} &= \frac{q}{h_2 - h_1} = \frac{755}{1508 - 98.5} \\ &= \boxed{0.536 \text{ lbm/sec}} \end{aligned}$$

The increase in power output is

$$h_2 - h_3' = 1508 - 1075 = 433 \text{ BTU/lbm}$$

Assume 100% mechanical efficiency.

$$\begin{aligned} P_{\text{steam turbine}} &= \frac{(0.536)(433)(778)}{550} \\ &= \boxed{328.3 \text{ hp}} \end{aligned}$$

2. Start with the new conditions.

At 1:

$$\begin{aligned} T_1 &= 900°\text{F} \\ p_1 &= 1000 \text{ psia} \\ h_1 &= 1448 \text{ BTU/lbm} \end{aligned}$$

At 2:

$$\begin{aligned} p_2 &= 40 \text{ psia} \\ h_2 &= 1122 \text{ BTU/lbm} \quad \text{[from Mollier diagram]} \\ h_2' &= 1448 - (0.90)(1448 - 1122) \\ &= 1155 \text{ BTU/lbm} \end{aligned}$$

At 3:

$$\begin{aligned} p_3 &= p_2 = 40 \text{ psia} \\ T_3 &= 900°\text{F} \\ h_3 &= 1481.4 \text{ BTU/lbm} \end{aligned}$$

At 4:

$$\begin{aligned} p_4 &= 2 \text{ psia} \\ h_4 &= 1156 \text{ BTU/lbm} \quad \text{[from Mollier diagram]} \\ h_4' &= 1481 - (0.80)(1481 - 1156) \\ &= 1221 \text{ BTU/lbm} \end{aligned}$$

At 5:

Assume the steam is saturated.

$$\begin{aligned} p_5 &= 2 \text{ psia} \\ T_5 &= 126.08°\text{F} \\ h_5 &= 94 \text{ BTU/lbm} \end{aligned}$$

At 6:

$$h_6 = h_5 = 94 \text{ BTU/lbm}$$

At 7:

$$\begin{aligned} p_7 &= 30 \text{ psia} \\ h_7 &= 219 \text{ BTU/lbm} \\ \upsilon_{F,7} &= 0.01701 \text{ ft}^3/\text{lbm} \end{aligned}$$

At 8:

$$\begin{aligned} p_8 &= 1000 \text{ psia} \\ h_8' &= 219 + \frac{(0.01701)(1000 - 30)(144)}{(778)(0.60)} \\ &= 224 \text{ BTU/lbm} \end{aligned}$$

At 9:

$$\begin{aligned} p_9 &= 70 \text{ psia} \\ h_9 &= 1165 \text{ BTU/lbm} \quad \text{[from Mollier diagram]} \\ h_9' &= 1448 - (0.90)(1448 - 1165) \\ &= 1193 \text{ BTU/lbm} \end{aligned}$$

The bleed fraction is unknown. Let x be the fraction.

$$\begin{aligned} x h_9 + (1 - x) h_6 &= h_7 \\ x(1193) + (1 - x)(94) &= 219 \\ x &= 0.114 \end{aligned}$$

Now, work with the original conditions.

The conditions at 1, 2, and 3 are unchanged.

At 4:

$$p_4 = 1 \text{ psia}$$
$$h_4 = 1109 \text{ BTU/lbm}$$
$$h'_4 = 1481.4 - (0.80)(1481.4 - 1109) = 1183 \text{ BTU/lbm}$$

At 5:

$$p_5 = 1 \text{ psia}$$
$$T_5 = 101.74°\text{F}$$
$$h_5 = 69.70 \text{ BTU/lbm}$$

At 6:

$$h_6 = h_5 = 69.70 \text{ BTU/lbm}$$

There are no changes in conditions at points 7, 8, and 9.

The bleed fraction must change to maintain the conditions leaving the heater.

$$xh_9 + (1-x)h_6 = h_7$$
$$x(1193) + (1-x)(69.70) = 219$$
$$x = 0.133$$

The net work is

$$\begin{aligned} W_{\text{net}} &= (h_1 - h'_9) \\ &\quad + (1-x)[(h'_9 - h'_2) + (h_3 - h'_4)] - (h_8 - h_7) \\ &= 1448 - 1193 \\ &\quad + (1 - 0.114)[(1193 - 1155) + (1481.4 - 1221)] \\ &\quad - (224 - 219) \\ &= 514.4 \text{ BTU/lbm} \end{aligned}$$

The original net work was

$$\begin{aligned} W_{\text{net}} &= 1448 - 1193 \\ &\quad + (1 - 0.133)[(1193 - 1155) + (1481.4 - 1183)] \\ &\quad - (224 - 219) \\ &= 541.7 \text{ BTU/lbm} \end{aligned}$$

$$\Delta W = 514.4 - 541.7 = -27.3 \text{ BTU/lbm} \quad \text{[decrease]}$$

$$\% = \frac{\Delta W}{W_{\text{original}}} = \frac{-27.3}{541.7} = -0.050 \quad \boxed{5\% \text{ decrease}}$$

3. Assume an air-standard gas turbine cycle. Use the air table because c_p is not constant over wide temperature variations.

At 1:

$$T_1 = 500°\text{R}$$
$$p_1 = 14.7 \text{ psia}$$
$$h_1 = 119.48 \text{ BTU/lbm}$$
$$p_r = 1.0590$$

At 2:

Compression ratios for turbines are pressure ratios.

$$p_2 = (10)(p_1) = (10)(14.7) = 147 \text{ psia}$$
$$p_r = (10)(1.0590) = 10.590$$
$$T_2 \approx 960°\text{R} \quad \text{[from the } p_r \text{ column]}$$
$$h_2 = 231.06 \text{ BTU/lbm}$$
$$h'_2 = 119.48 + \frac{231.06 - 119.48}{0.85} = 250.75 \text{ BTU/lbm}$$
$$T'_2 = 1040°\text{R} \quad \text{[from } h \text{ column]}$$

At 3:

$$p_3 = 147 \text{ psia}$$
$$T_3 = 2400°\text{R}$$
$$h_3 = 617.22 \text{ BTU/lbm}$$
$$p_{r,3} = 367.6$$

At 4:

$$p_{r,4} = (367.6)\left(\frac{18}{147}\right) = 45.0$$
$$T_4 = 1418°\text{R}$$
$$h_4 = 347.61 \text{ BTU/lbm}$$
$$h'_4 = 617.22 - (0.65)(617.22 - 347.61) = 442.0 \text{ BTU/lbm}$$
$$T'_4 = 1772°\text{R}$$

At 5:

$$T_5 = 760°\text{R}$$
$$h_5 = 182.08 \text{ BTU/lbm}$$

At 6:

$$T_6 = 1460°\text{R} = 1000°\text{F}$$
$$p_6 = 300 \text{ psia}$$
$$h_6 = 1525.2 \text{ BTU/lbm}$$

At 7:

$$p_7 = 10 \text{ psia}$$
$$h_7 = 1146 \text{ BTU/lbm}$$
$$h'_7 = 1525.2 - (0.80)(1525.2 - 1146) = 1222 \text{ BTU/lbm}$$

At 8:

Assume the steam is saturated.

$$p_8 = 10 \text{ psia}$$
$$T_8 = 193.21°\text{R}$$
$$h_8 = 161 \text{ BTU/lbm}$$
$$v_{F,8} = 0.01659 \text{ ft}^3/\text{lbm}$$

At 9:

$$p_9 = p_6 = 300 \text{ psia}$$
$$h'_9 = 161 + \frac{(0.01659)(300 - 10)(144)}{(778)(0.60)} = 162.5 \text{ BTU/lbm}$$

Determine the gas turbine load in BTU/hr.

$$\frac{5000 \text{ kW}}{(0.97)(0.293 \times 10^{-3})} = 1.76 \times 10^7 \text{ BTU/hr [net]}$$

The net work per pound of gas is

$$(h_3 - h'_4) - (h'_2 - h_1) = (617.22 - 442.0) - (250.75 - 119.48) = 43.95 \text{ BTU/lbm}$$

The total air flow through the turbine is

$$\frac{1.76 \times 10^7 \ \frac{\text{BTU}}{\text{hr}}}{43.95 \ \frac{\text{BTU}}{\text{lbm}}} = 400{,}455 \text{ lbm/hr}$$

Neglecting the fuel mass, the energy entering the gases in the combustor is

$$(400{,}455)(617.22 - 250.75) = 1.47 \times 10^8 \text{ BTU/hr}$$

The combustion energy is

$$\left(8500 \ \frac{\text{lbm}}{\text{hr}}\right)\left(18{,}000 \ \frac{\text{BTU}}{\text{lbm}}\right) = 1.53 \times 10^8 \text{ BTU/hr}$$

The combustor efficiency is

$$\frac{1.47 \times 10^8}{1.53 \times 10^8} = \boxed{0.96 \quad \text{[seems high]}}$$

The steam flow rate is

$$\dot{m}_{\text{steam}} = \frac{\eta \dot{m}_4 (h'_4 - h_5)}{h_6 - h'_9} = \frac{(0.85)(400{,}455)(442.0 - 182.08)}{1525.2 - 162.5} = 6.49 \times 10^4 \text{ lbm/hr}$$

Neglecting the feed pump work, the steam turbine output is

$$(6.49 \times 10^4)(1525 - 1222)(0.293 \times 10^{-3}) = \boxed{5762 \text{ kW}}$$

The overall efficiency is

$$\eta = \frac{W_{\text{net}}}{Q_{\text{in}}} = \frac{5000 \text{ kW} + 5762 \text{ kW}}{(1.53 \times 10^8)(0.293 \times 10^{-3})} = \boxed{0.24 \ (24\%)}$$

4. Noncondensables are gases (hydrogen and carbon dioxide) in solution with boiler feed water liberated in the boiler and also air that has leaked through seals.

Noncondensables are undesirable because they decrease the pressure to which steam expands (if allowed to build up) and contribute to corrosion.

Gases are removed by steam jet ejectors (air ejectors) and positive displacement pumps.

(a)

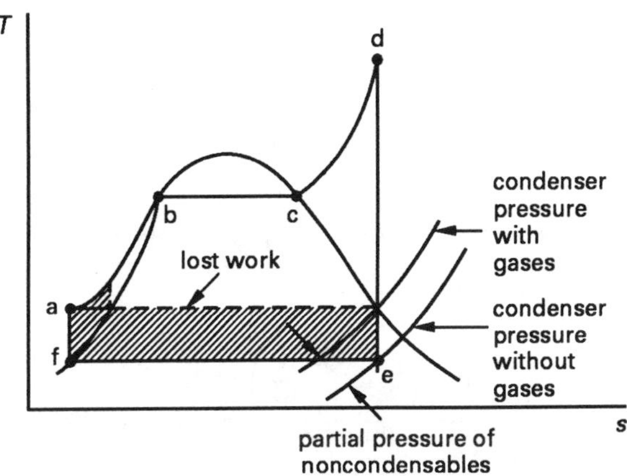

(b) $p_{\text{absolute}} = 30 - 27 = 3 \text{ in Hg}$

$$= \left(0.491 \ \frac{\text{lbf}}{\text{in}^3}\right)(3 \text{ in}) = 1.473 \text{ psia}$$

From Dalton's Law,

$$p_{\text{total}} = p_{\text{steam}} + p_{\text{nc}}$$

At 80°F, $p_{\text{steam}} = 0.5069$ psi, so

$$p_{\text{nc}} = 1.473 - 0.5069 = 0.9661 \text{ psi}$$

Assume noncondensables are air ($R = 53.3$) and that water vapor is a perfect gas ($R = 85.8$).

From $pV = mRT$,

$$m = pV/RT$$

V and T are the same for both the gases and steam, so

$$m_{nc} \propto \frac{p}{R} = \frac{0.9961}{53.3} = 0.01813$$
$$m_{steam} \propto \frac{p}{R} = \frac{0.5069}{85.8} = 0.005908$$

The gravimetric fraction of the noncondensables is

$$G_{nc} = \frac{0.01813}{0.01813 + 0.005908} = \boxed{0.754\ (75.4\%)}$$

5. Based on the test data,

$$0 = (4.3 \times 10^{-3})(65)[(0.016)(2100) + (F)(20)(65)^2 - (0.07)(2100)]$$

The quantity in brackets must be zero. The streamline factor is

$$F = 1.342 \times 10^{-3}$$

(a) $$(1.20)\left[(4.3 \times 10^{-3})(55)[(0.016)(2100) + (1.342 \times 10^{-3})(20)(55)^2 + (0.06)(2100)]\right] = \boxed{68.3\ \text{hp}}$$

(b) The fuel used is

$$\frac{(68.3\ \text{hp})\left(0.43\ \frac{\text{lbm}}{\text{hp-hr}}\right)}{5.8\ \frac{\text{lbm}}{\text{gal}}} = 5.06\ \text{gal/hr}$$

The mileage is

$$\frac{55\ \frac{\text{mi}}{\text{hr}}}{5.06\ \frac{\text{gal}}{\text{hr}}} = \boxed{10.9\ \text{mi/gal}}$$

6. Notice the pressures are in psig.

(a) Entering the turbine:

$$p_1 = 800 + 14.7 \approx 815\ \text{psia}$$
$$h_1 \approx 1455\ \text{BTU/lbm} \quad \text{[from Mollier diagram]}$$

In the condenser:

$$h_2 \approx 915\ \text{BTU/lbm} \quad \text{[from Mollier diagram]}$$
$$h_2' = 1455 - (0.80)(1455 - 915) = \boxed{1023\ \text{BTU/lbm} \quad \text{[same for both options]}}$$

(b) Option 1:

$$\dot{m} = 170{,}000 - 55{,}000 = 115{,}000\ \text{lbm/hr}$$
$$P = (115{,}000)(1455 - 1023)(0.98)(0.99)(0.293 \times 10^{-6}) = 14.1\ \text{MW}$$

Option 2:

$$p_4 = 120 + 14.7 \approx 135\ \text{psia}$$
$$h_4 \approx 1240\ \text{BTU/lbm} \quad \text{[from Mollier diagram]}$$
$$h_4' = 1455 - (0.80)(1455 - 1240) = 1283\ \text{BTU/lbm}$$
$$P = (0.98)(0.99)(0.293 \times 10^{-6}) \times [(170{,}000)(1455 - 1283) + (115{,}000)(1283 - 1023)] = 16.8\ \text{MW}$$

$$\Delta P = 16.8 - 14.1 = \boxed{2.7\ \text{MW}}$$

7. R-22 data and a p-h diagram are needed.

Evaporator:

For heat to flow from the cooled environment, the refrigerant in the evaporator must be cooler. During phase changes, the difference between fluid and environment is typically 5–30°F. In this case, say 20°F, so liquid R-22 enters at −70°F.

Intermediate exchanger:

Use the $\Delta T = 20$°F again as the average temperature difference between R-22 and R-12. The R-22 enters the intermediate exchanger as a superheated vapor. Assuming type "L" copper tubing and a maximum working pressure of 250 psia, this sets the maximum temperature for R-22 at 240°F.

R-12 refrigerators generally operate with a 40–45°F evaporation temperature, so choose 40°F as the R-12 temperature in the intermediate exchanger. This sets the R-22 saturation temperature in the exchanger as 40°F + 20°F = 60°F.

Condenser:

Again, use $\Delta T = 20$°F. The temperature for the R-12 would be 85°F + 20°F = 105°F.

The T-s diagram for the processes would be

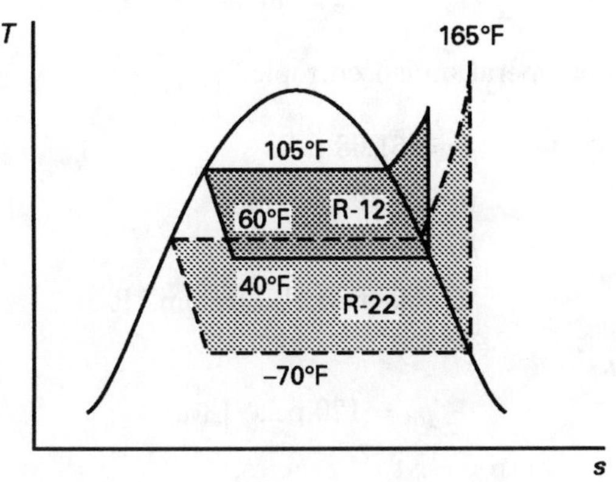

Evaporator:

The pressure in the evaporator is $p_{\text{saturated}}$ corresponding to −70°F.

6.6 psia [say 7 psia]

Intermediate exchanger:

R-22: Change of phase occurs at the saturation pressure corresponding to 60°F.

approximately 116 psia

R-12: The pressure corresponding to 40°F is

52 psia

Condenser:

The superheat temperature is unknown, but following similar reasoning, the entering pressure would be kept below 250 psia.

Change of phase takes place at a pressure corresponding to 105°F.

140 psia

8. At 1:

$$T_1 = 460 + 60 = 520°\text{R}$$
$$h_1 = 124.27 \text{ BTU/lbm}$$
$$p_{r,1} = 1.2147$$

At 2:

$$p_2 = 150 \text{ psia}$$

If isentropic,

$$p_{r,2} = \left(\frac{150}{14.7}\right)(1.2147) = 12.395$$

From the air table, given $p_{r,2}$,

$$T_2 = 1002°\text{R}$$
$$h_2 = 241.48 \text{ BTU/lbm}$$

But, $\eta = 0.85$.

$$\begin{aligned} h_2' &= 124.27 + \frac{241.48 - 124.27}{0.85} \\ &= 262.16 \text{ BTU/lbm} \end{aligned}$$
$$T_2' = 1085°\text{R} \quad \text{[from the air table given } h_2'\text{]}$$
$$\begin{aligned} W_{\text{compression}} &= h_2' - h_1 \\ &= 262.16 - 124.27 \\ &= 137.89 \text{ BTU/lbm} \end{aligned}$$

At 3:

$$T_3 = 2240 + 460 = 2700°\text{R}$$
$$h_3 = 703.35 \text{ BTU/lbm}$$
$$p_{r,3} = 601.9$$
$$\begin{aligned} q_{\text{in}} &= h_3 - h_2' \\ &= 703.35 - 262.16 \\ &= 441.19 \text{ BTU/lbm} \end{aligned}$$

At 4:

$$p_4 = \frac{150}{10} = 15 \text{ psia}$$

If isentropic,

$$p_{r,4} = \frac{601.9}{10} = 60.19$$
$$T_4 = 1531°\text{R}$$
$$h_4 = 377.46 \text{ BTU/lbm}$$

But, $\eta = 0.74$.

$$\begin{aligned} h_4' &= 703.35 - (0.74)(703.35 - 377.46) \\ &= 462.19 \text{ BTU/lbm} \end{aligned}$$
$$\begin{aligned} W_{\text{turbine}} &= h_3 - h_4' \\ &= 703.35 - 462.19 \\ &= 241.16 \text{ BTU/lbm} \end{aligned}$$

At 5:

$$T_5 = 400 + 460 = 860°\text{R}$$
$$h_5 = 206.46 \text{ BTU/lbm}$$

At 6:

$$T_6 = 250°\text{R}$$
$$h_6 = 218.48 \text{ BTU/lbm}$$

At 7:

$$h_7 = 1248.6 \text{ BTU/lbm}$$
$$s_7 = 1.6756 \text{ BTU/lbm-°R}$$

At 8:

$$p_8 = 30 \text{ psia}$$

If isentropic, $s_8 = s_7$.

$$x = \frac{s - s_F}{s_{FG}} = \frac{1.6756 - 0.3680}{1.3313} = 0.982$$
$$\begin{aligned} h_8 &= h_F + xh_{FG} \\ &= 218.82 + (0.982)(945.3) \\ &= 1147.1 \text{ BTU/lbm} \end{aligned}$$

But $\eta = 0.82$.

$$\begin{aligned} h_8' &= 1248.6 - (0.82)(1248.6 - 1147.1) \\ &= 1165.4 \text{ BTU/lbm} \end{aligned}$$
$$\begin{aligned} W_{\text{steam turbine}} &= h_7 - h_8' = 1248.6 - 1165.4 \\ &= 83.2 \text{ BTU/lbm} \end{aligned}$$

(b) The energy given up by gas in the waste heat boiler is

$$\begin{aligned} q_{\text{out}} &= h_4' - h_5 \\ &= 462.19 - 206.46 \\ &= 255.73 \text{ BTU/lbm} \end{aligned}$$

The energy absorbed by steam in the waste heat boiler is

$$\begin{aligned} q_{\text{in}} &= h_7 - h_6 = 1248.6 - 218.48 \\ &= 1030.12 \text{ BTU/lbm} \end{aligned}$$

Since $Q_{\text{in}} = m_{\text{steam}}q_{\text{in}} = Q_{\text{out}} = m_{\text{gas}}q_{\text{out}}$, the ratio of mass flows is

$$\frac{\dot{m}_{\text{steam}}}{\dot{m}_{\text{gas}}} = \frac{255.73}{1030.12} = \boxed{0.248}$$

(a)
$$\eta = \frac{W_{\text{out}} - W_{\text{in}}}{Q_{\text{in}}} \quad \text{[negligible pump work]}$$
$$= \frac{241.16 + (0.248)(83.2) - 137.89}{441.19}$$
$$= \boxed{0.281\ (28.1\%)}$$

9.

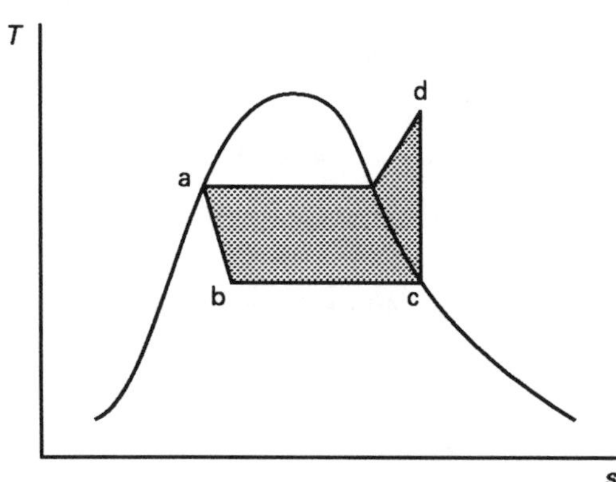

At c (entering the compressor):

$$T_c = 40\text{°F}$$

Assume operation is isentropic.

$$p = 51.68 \text{ psia}$$
$$v_c = 0.792 \text{ ft}^3\text{/lbm}$$
$$h_c = 82.71 \text{ BTU/lbm}$$
$$s_c = 0.16833 \text{ BTU/lbm-°R}$$

At d:

$$p_d = 120 \text{ psia} \quad \text{[given]}$$

For an isentropic c-to-d process,

$$s_d = s_c = 0.16833 \text{ BTU/lbm-°R}$$

Searching the superheat table at $p_d = 120$ psia and $s_d \approx 0.16833$,

$$T_d \approx 100\text{°F}$$

(Ideal gas formulas cannot be used to calculate T_d since this is a vapor.)

$$h_d \approx 89.13 \text{ BTU/lbm}$$
$$h_d' = 82.71 + \frac{89.13 - 82.71}{0.60} = 93.41 \text{ BTU/lbm}$$

At a:

$$p_a = 120 \text{ psia}$$

Assume saturated.

$$T_a = 93.4\text{°F}$$
$$h_a = 29.53 \text{ BTU/lbm}$$

At b:

$$T_b = 40\text{°F}$$
$$p_b = 51.68 \text{ psia}$$
$$h_b = h_a = 29.53 \text{ BTU/lbm}$$

The refrigeration effect is between b and c.

$$\dot{m} = \frac{(8 \text{ tons})\left(12{,}000 \ \frac{\text{BTU}}{\text{hr-ton}}\right)}{82.7 \ \frac{\text{BTU}}{\text{lbm}} - 29.53 \ \frac{\text{BTU}}{\text{lbm}}} = 1806 \text{ lbm/hr}$$

(b) Compressor work is between c and d.

$$P = (1806)(93.41 - 82.71)(0.293 \times 10^{-3})$$
$$= \boxed{5.66 \text{ kW}}$$

Notice that the efficiency was used to calculate h_d and should not be duplicated here.

The volumetric efficiency is

$$\eta_v = 1 - \left[\left(\frac{120}{51.68}\right)^{\frac{1}{1.3}} - 1\right](0.06) = 0.945$$

(a) The piston displacement (swept volume) is

$$\frac{\text{displacement}}{\text{stroke}} = \frac{\left(1806\ \frac{\text{lbm}}{\text{hr}}\right)\left(0.792\ \frac{\text{ft}^3}{\text{lbm}}\right)}{(0.945)\left(60\ \frac{\text{min}}{\text{hr}}\right)\left(125\ \frac{\text{strokes}}{\text{min}}\right)} = \boxed{0.20\ \text{ft}^3}$$

10. Noncondensing:

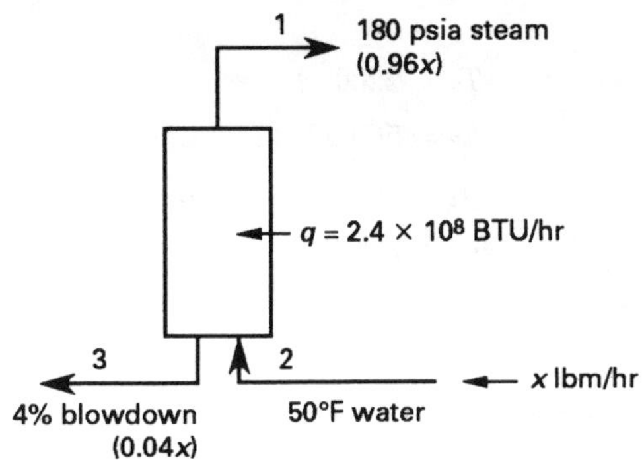

The answer will vary with assumed enthalpy of the blowdown liquid. Assume blowdown is 180 psia saturated liquid.

Assume boiler efficiency is 100%. (This is not realistic, but is done to allow uniform comparison with other solutions—it should actually be 85–90%.)

Assume feedwater and condensate pump work can be neglected.

$$h_1 = 1196.9\ \text{BTU/lbm}$$
$$h_2 = 18.07\ \text{BTU/lbm}$$
$$h_3 = 346.03\ \text{BTU/lbm}$$

Let x = number of pounds of 50°F water supplied per hour. The energy balance is

$$2.4 \times 10^8 = (1 - 0.04)(x)(1196.9 - 18.07) + (0.04)(x)(346.03 - 18.07)$$
$$x = 2.10 \times 10^5\ \text{lbm/hr of water}$$

The steam production is

$$(1 - 0.04)(2.10 \times 10^5) = \boxed{2.02 \times 10^5\ \text{lbm/hr}}$$

Condensing:

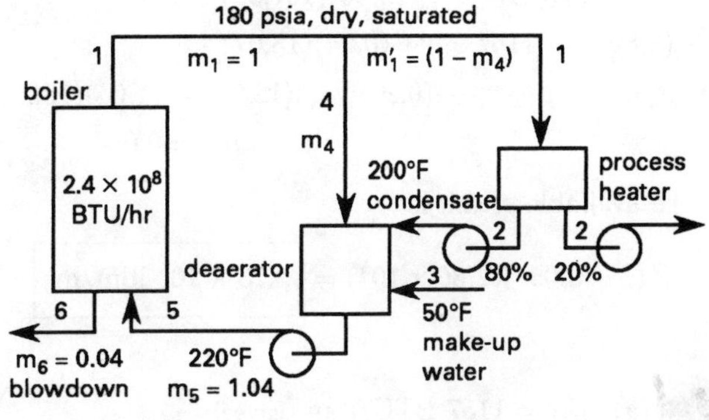

$$h_1 = 1196.9\ \text{BTU/lbm}$$
$$h_2 = 167.99\ \text{BTU/lbm}$$
$$h_3 = 18.07\ \text{BTU/lbm}$$
$$h_4 = h_1 = 1196.9\ \text{BTU/lbm}$$
$$h_5 = 188.13\ \text{BTU/lbm}$$
$$h_6 = 346.03\ \text{BTU/lbm}$$

The energy balance in the boiler is

$$2.4 \times 10^8 = (0.96)(x)(1196.9 - 188.13) + (0.04)(x)(346.03 - 188.13)$$
$$x = \text{incoming 220°F water} = 2.46 \times 10^5\ \text{lbm/hr}$$

The steam production is

$$(0.96)x = (0.96)(2.46 \times 10^5) = 2.36 \times 10^5\ \text{lbm/hr}$$

This is the steam produced, not the steam available for process heating.

The mass flow rates are

$$m_1 = 1$$
$$m_1' = 1 - m_4$$
$$m_2 = (0.80)(1 - m_4)$$
$$m_2' = (0.20)(1 - m_4)$$
$$m_3 = m_2' - m_6$$

m_4 is unknown.

$$m_5 = \frac{1}{1 - 0.04} = 1.042\ \ \text{[say 1.04]}$$
$$m_6 = (0.04)m_5 = (0.04)(1.04) = 0.042\ \ \text{[say 0.04]}$$

The energy balance in the deaerator is

$$m_4h_4 + m_2h_2 + m_3h_3 = m_5h_5$$
$$m_4h_4 + (0.80)(1 - m_4)h_2 + [(0.20)(1 - m_4) + 0.04]h_3 = (1.04)h_5$$

Substituting values,

$$m_4(1196.9) + (0.80)(167.99) - (0.80)(m_4)(167.99) + (0.24)(18.07) - (0.20)(m_4)(18.07) = (1.04)(188.13)$$

$$m_4 = 0.054$$

The available steam is

$$(1 - 0.054)(2.36 \times 10^5) = \boxed{2.23 \times 10^5 \text{ lbm/hr}}$$

11. (a)

$$h_1 = 1157 \text{ BTU/lbm [given]}$$
$$h_2 = 188.13 \text{ BTU/lbm}$$
$$h_3 = 157.95 \text{ BTU/lbm}$$
$$h_4 = \text{unknown}$$
$$m_1 = 1.00 - 0.14 - 0.80 = 0.06 \text{ [millions]}$$

The energy balance is

$$m_1h_1 + m_2h_2 + m_3h_3 = m_4h_4$$

$$(0.06)(1157) + (0.80)(188.13) + (0.14)(157.95) = (1.00)h_4$$

$$h_4 = 242.04 \text{ BTU/lbm}$$

Searching the steam tables for a pressure with h_F = 242.04 BTU/lbm,

$$p \approx \boxed{44 \text{ psia}}$$

Since hot well depression is zero, the condensate is not subcooled.

(b) For 44 psia steam,

$$x = \frac{h - h_F}{h_{FG}} = \frac{1157 - 241.95}{929.6} = \boxed{0.984}$$

12.

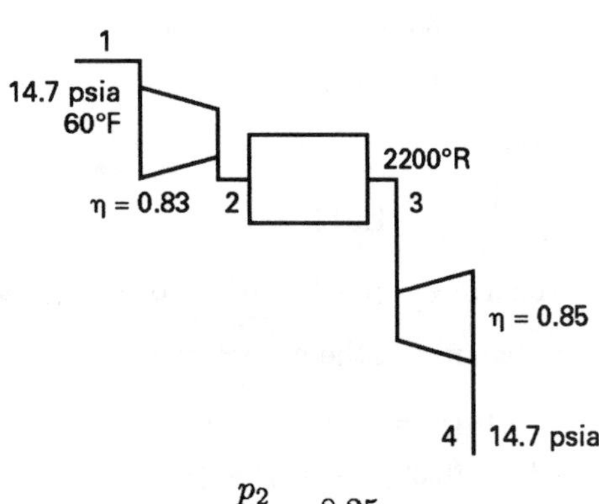

$$\frac{p_2}{p_1} = 8.25$$

As is:

$$m = 1$$

At 1:

$$T_1 = 60°\text{F} = 520°\text{R}$$
$$h_1 = 124.27 \text{ BTU/lbm}$$
$$p_{r,1} = 1.2147$$

At 2:

$$p_2 = (14.7)(8.25) = 121.3 \text{ psia}$$
$$p_{r,2} = (1.2147)(8.25) = 10.02$$
$$T_2 = 945°\text{R}$$
$$h_2 = 227.35 \text{ BTU/lbm}$$
$$h_2' = 124.27 + \frac{227.35 - 124.27}{0.83} = 248.46 \text{ BTU/lbm}$$

At 3:

$$T_3 = 2200°\text{R}$$
$$h_3 = 560.59 \text{ BTU/lbm}$$
$$p_3 = p_2 = 121.3 \text{ psia}$$
$$p_{r,3} = 256.6$$

At 4:

$$p_4 = 14.7 \text{ psia}$$
$$p_{r,4} = p_{r,3}\left(\frac{p_4}{p_3}\right) = (256.6)\left(\frac{1}{8.25}\right) = 31.1$$
$$T_4 = 1286°\text{R}$$
$$h_4 = 313.33 \text{ BTU/lbm}$$
$$h_4' = 560.59 - (0.85)(560.59 - 313.33) = 350.4 \text{ BTU/lbm}$$
$$W_{\text{net}} = W_{\text{turbine}} - W_{\text{compressor}} = m\Delta h = (560.59 - 350.4) - (248.46 - 124.27) = 86.0 \text{ BTU/lbm}$$

With bypass:

$$m = 0.95$$

At 1:

$$T_1 = 520°\text{R}$$
$$h_1 = 124.27 \text{ BTU/lbm}$$
$$p_{r,1} = 1.2147$$

At 2:

$$p_2 = 121.3 + 5 = 126.3 \text{ psia}$$

$$p_{r,2} = (1.2147)\left(\frac{126.3}{14.7}\right) = 10.437$$

$$T_2 = 956°\text{R}$$

$$h_2 = 230.07 \text{ BTU/lbm}$$

$$h_2' = 124.27 + \frac{230.07 - 124.27}{0.83} = 251.74 \text{ BTU/lbm}$$

At 3:

$$T_3 = 2450°\text{R}$$

$$h_3 = 631.48 \text{ BTU/lbm}$$

$$p_3 = p_2 = 126.3 \text{ psia}$$

$$p_{r,3} = 400.5$$

At 4:

$$p_4 = 14.7 \text{ psia}$$

$$p_{r,4} = (400.5)\left(\frac{14.7}{126.3}\right) = 46.6$$

$$T_4 = 1431°\text{R}$$

$$h_4 = 351.01 \text{ BTU/lbm}$$

$$h_4' = 631.48 - (0.85)(631.48 - 351.01) = 393.08 \text{ BTU/lbm}$$

$$W = (0.95)[(631.48 - 393.08) - (251.74 - 124.27)] = 105.4 \text{ BTU/lbm}$$

Since 105.4 > 86.0, the engine will produce more thrust.

yes

COMPRESSIBLE FLUID FLOW

1. (a) $p_e/p_o = 0.1278$ corresponds to $M = 2$.

At $M = 2$,

$$\frac{A_e}{A_t} = 1.6875$$

A shock wave at the exit represents case E (i.e., second critical pressure ratio).

$$p_{x,\mathrm{E}} = p_G = (0.1278)p_o$$
$$M_{x,\mathrm{E}} = M_G = 2$$

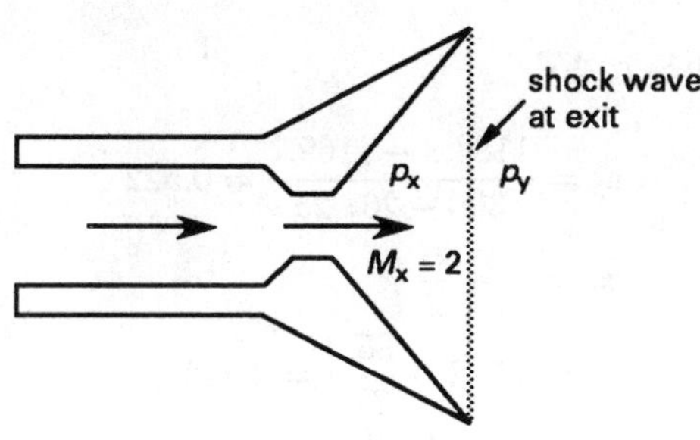

From the normal shock tables for $M = 2$,

$$\frac{p_y}{p_x} = 4.5$$
$$p_y = 4.5p_x$$
$$= (4.5)(0.1278)p_o$$

Then,

$$\frac{p_y}{p_o} = 0.5751$$

(b) At $M = 2$, $A_e/A_t = 1.6875$. Since $A = \frac{\pi}{4}D^2$,

$$\frac{D_e}{D_t} = \sqrt{1.6875} = 1.3$$

The nozzle looks like this:

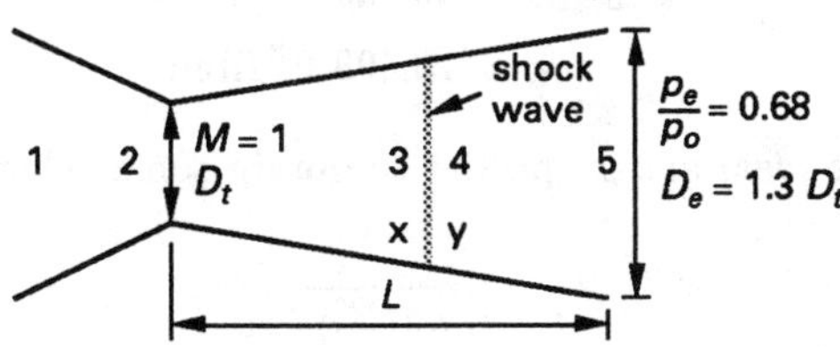

$p_{T,5}$ is not known, so M_5 cannot be initially determined.

Flow is subsonic since $p_e/p_o > 0.5283$ at exit.

However, $p_e = 0.68p_o$. (p_e is a static pressure.)

Use the following procedure.

First iteration:

step 1: Assume $M_x = 1.7$ at shock.

step 2: Read $p_{T,4}/p_{T,3} = 0.8557$ from $p_{T,y}/p_{T,x}$ column for $M = 1.7$.

step 3: Flow is isentropic from 1 to 3, so $p_{T,1} = p_{T,3}$. In terms of table factors,

$$\left[\frac{p_x}{p_{T,y}}\right] = \left[\frac{p_x}{p_{T,x}}\right]\left[\frac{p_{T,x}}{p_{T,y}}\right]$$

In terms of nozzle locations, since $p_{T,1} = p_{T,3}$,

$$\frac{p_5}{p_{T,4}} = \left(\frac{p_5}{p_{T,1}}\right)\left(\frac{p_{T,3}}{p_{T,4}}\right) \quad [\text{since } p_o = p_{T,3}]$$
$$\frac{p_5}{p_{T,4}} = \frac{0.68p_o}{0.8557p_{T,3}} = \frac{0.68}{0.8557}$$
$$= 0.7947$$

step 4: Flow is isentropic between 4 and 5, so $p_{T,4} = p_{T,5}$. Find M_5.

$$\frac{p}{p_T} = \frac{p_5}{p_{T,4}} = \frac{p_5}{p_{T,5}} = 0.7947$$

Reading from the table for $p/p_T = 0.7947$, $M_{\text{exit}} = M_5 \approx 0.58$, and the exit area ratio should be

$$\frac{A_e}{A_y^*} = 1.2130$$

Note: A_x^* and A_y^* are not the same.

step 5: For the actual nozzle, $A_e/A_t = 1.6875$, so

$$\frac{A_e}{A_y^*} = (1.6875)(0.8557)$$
$$= 1.44$$

Since $1.44 \neq 1.2130$, $M_x \neq 1.7$.

Additional iterations are needed to find M_x. Subsequent iterations are summarized here.

	iteration		
	1	2	3
assumed M_x	1.7	1.8	1.83
$\frac{p_{T,4}}{p_{T,5}}$ read as $\left[\frac{p_{T,y}}{p_{T,x}}\right]$	0.8557	0.8127	0.7993
$\frac{p_5}{p_{T,4}} = \frac{0.68}{\left[\frac{p_{T,4}}{p_{T,5}}\right]}$	0.7947	0.837	0.8507
M_{exit} from the table	0.58	0.51	0.48
$\frac{A_{\text{exit}}}{A^*}$ from table	1.213	1.321	1.3595
$\frac{A_{\text{exit}}}{A^*} = (1.6875)\left(\frac{p_{T,4}}{p_{T,5}}\right)$	1.44	1.37	1.35 [close enough]

$M_x \approx 1.83$ and $M_{\text{exit}} \approx 0.48$. For $M = 1.83$,

$$\frac{A_{\text{shock}}}{A_t} = 1.472$$

$$\frac{D_{\text{shock}}}{D_t} = \sqrt{1.472} = 1.21$$

Distance from shock to throat is

$$\left(\frac{1.21 - 1.00}{1.3 - 1.00}\right) L = \boxed{0.70L}$$

2. (a) Assume metabolic heats of

$$q_s = 250 \text{ BTU/hr-person}$$

$$q_l = 200 \text{ BTU/hr-person}$$

Assume $h = 1054$ BTU/lbm for 70°F. From the psychrometric chart,

$$\omega_1 = 44 \text{ grains/lbm } [0.00629 \text{ lbm/lbm}]$$

From a low-temperature psychrometric chart,

$$\omega_o = 0.00016 \text{ lbm/lbm}$$

$$v_o = 11.95 \text{ ft}^3\text{/lbm}$$

The moisture losses and gains are

$$\dot{m}_{w,\text{people}} = \frac{(6)(200)}{1054} \approx 1.1 \text{ lbm/hr}$$

$$\dot{m}_{w,\text{beets}} = 350 \text{ lbm/hr}$$

$$\begin{aligned} \dot{m}_{w,\text{infiltration}} &= \dot{m}_a \Delta\omega \\ &= (25{,}000)(60)\left(\frac{1}{11.95}\right) \\ &\quad \times (0.00629 - 0.00016) \\ &= 769.5 \text{ lbm/hr} \end{aligned}$$

The total steam needs are

$$350 + 769.5 - 1.1 = \boxed{1118.4 \text{ lbm/hr}}$$

(b) Assume $k_{\text{steam}} = 1.28$.

Proof:

$$p = 25 \text{ psig} + 14.7 \approx 40 \text{ psia}$$

At 40 psia (saturated),

$$T_{\text{saturated}} = 267.25°\text{F}$$

$$h = 1169.7 \text{ BTU/lbm}$$

At 40 psia and 300°F (superheated),

$$h = 1186.8 \text{ BTU/lbm}$$

Since $\Delta h = c_p \Delta T$,

$$c_p = \frac{1186.8 - 1169.7}{300 - 267.25} = 0.522$$

$$\begin{aligned} c_v &= c_p - \frac{R}{J} \\ &= 0.522 - \frac{85.8}{778} = 0.412 \end{aligned}$$

$$k = \frac{c_p}{c_v} = \frac{0.522}{0.412} = 1.27$$

It would be acceptable to assume $k = 1.30$ and use tables, but this solution does not.

The critical pressure ratio is

$$\begin{aligned} R_{cp} &= \left(\frac{2}{1.28 + 1}\right)^{\frac{1.28}{1.28-1}} \\ &= 0.55 \end{aligned}$$

The steam pressure is $25 + 14.7 \approx 40$.

$$\frac{p_{\text{room}}}{p_{\text{steam}}} = \frac{14.7}{40} = 0.37$$

Since $0.37 < 0.55$, steam flow will be choked.

At 40 psia saturated,

$$T_{\text{saturated}} = 267.25°\text{F}$$

$$v_G = 10.498 \text{ ft}^3\text{/lbm}$$

Assume ideal gas properties to obtain sonic velocity at the throat.

$$c = \sqrt{kgRT_{\text{throat}}}$$

$$\begin{aligned} \left.\frac{T_{\text{total}}}{T_{\text{throat}}}\right|_{M=1} &= \left(\tfrac{1}{2}\right)(k-1)M^2 + 1 \\ &= \left(\tfrac{1}{2}\right)(1.28 - 1)(1)^2 + 1 \\ &= 1.14 \end{aligned}$$

$$c = \sqrt{\frac{(1.28)(32.2)(85.8)(460 + 267.25)}{1.14}}$$
$$= 1502.0 \text{ ft/sec}$$

The mass flow rate is

$$\dot{m} = \mathrm{v}A\rho$$

$$\frac{\rho_{\text{total}}}{\rho_{\text{throat}}} = \left(\frac{T_{\text{total}}}{T_{\text{throat}}}\right)^{\frac{1}{k-1}} = (1.14)^{\frac{1}{1.28-1}}$$
$$= 1.5967$$

$$\rho_{\text{throat}} = \left(\frac{1}{10.498}\right)\left(\frac{1}{1.5967}\right)$$
$$= 0.05966 \text{ lbm/ft}^3$$

The total required area is

$$A_{\text{required}} = \frac{\dot{m}}{\mathrm{v}_{\text{throat}}\rho_{\text{throat}}} = \frac{1118.4}{(3600)(1502)(0.05966)}$$
$$= 0.003467 \text{ ft}^2$$

The area per orifice is

$$A_{\text{orifice}} = \left(\frac{\pi}{4}\right)\left(\frac{0.37}{12}\right)^2 = 0.000747 \text{ ft}^2$$

The number of orifices needed is

$$n = \frac{0.003467}{0.000747} = \boxed{4.64 \quad \text{[say 5 orifices]}}$$

This calculation ignores a discharge coefficient, which could have been assumed and used.

3. Assumptions:

- ideal gas
- simple compressibility model
- orifice flow is one-dimensional and isentropic
- $C_d = 0.70$ (certainly much less than 1.0)

Calculate hydrogen mass. The reduced pressure and temperature are

$$p_r = \frac{1100}{188} = 5.85$$
$$T_r = \frac{600 + 460}{60.5} = 17.52$$

From a standard compressibility factor chart, $Z = 1.03$.

$$R = 766.8$$
$$k = 1.41 \quad \left[\begin{array}{c}\text{assume } k = 1.40 \text{ so that isentropic} \\ \text{flow tables can be used}\end{array}\right]$$
$$m = \frac{pV}{ZRT} = \frac{(1100)(144)(900)}{(1.03)(766.8)(1060)}$$
$$= 170.3$$

Calculate the flow rate.

The critical pressure ratio is

$$R_{\text{cp}} = \left(\frac{2}{1.40 + 1}\right)^{\frac{1.40}{1.40-1}} = 0.5283$$

In this problem,

$$\frac{p_{\text{outside}}}{p_{\text{inside}}} = \frac{14.7}{1100} = 0.013 < R_{\text{cp}}$$

Flow is choked; that is, flow is critical and sonic through the orifice.

$$A_{\text{orifice}} = \left(\frac{\pi}{4}\right)\left(\frac{0.5}{12}\right)^2 = 1.364 \times 10^{-3} \text{ ft}^2$$

From the isentropic flow tables, for $M = 1$,

$$\frac{p}{p_T} = 0.5283$$
$$\frac{T}{T_T} = 0.8333$$

The conditions of the gas in the orifice are

$$p = (0.5283)(1100) = 581.1 \text{ psia}$$
$$T = (0.8333)(1060) = 883.3^\circ\text{R}$$

The gas is flowing at the speed of sound in the orifice. The speed of sound is

$$c = \sqrt{kgRT} = \sqrt{(1.4)(32.2)(766.8)(883.3)}$$
$$= 5526 \text{ ft/sec}$$

At the orifice conditions, the reduced variables are

$$p_r = \frac{p}{p_c} = \frac{581.1}{188} = 3.1$$
$$T_r = \frac{T}{T_c} = \frac{883.3}{60.5} = 14.6$$

The compressibility factor is

$$Z = 1.02$$

The hydrogen density in the orifice is

$$\rho = \frac{p}{ZRT} = \frac{(581.1)(144)}{(1.02)(766.8)(883.3)}$$
$$= 0.1211 \text{ lbm/ft}^3$$

The mass flow rate is

$$\dot{m} = C_d \rho c A = (0.70)(0.1211)(5526)(1.364 \times 10^{-3})$$
$$= 0.639 \text{ lbm/sec}$$

Calculate the time.

$$t = \frac{m}{\dot{m}} = \frac{170.3}{0.639} = \boxed{267 \text{ sec}}$$

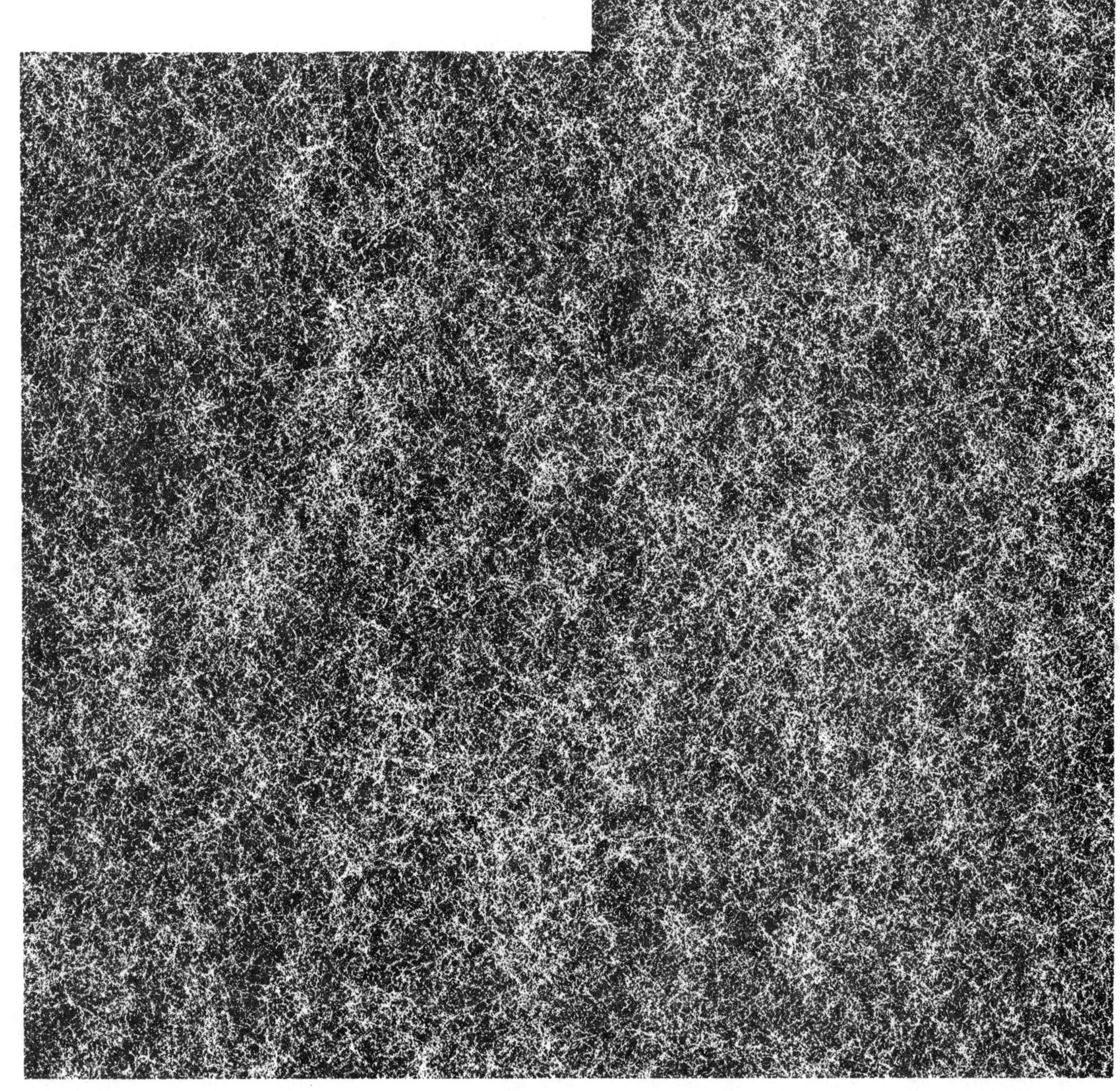

COMBUSTION

1. There is no sulfur in the fuel, so the only vapor condensing will be water vapor. (The Orsat analysis is dry and shows no water. However, there is water in the flue gas.)

Work with 100 ft^3 of flue gas. All of the nitrogen in the flue gas came from the combustion air. If 84.0 ft^3 of nitrogen enters the furnace, then the volume of oxygen entering the furnace is approximately

$$\frac{84.0\ \text{ft}^3}{3.78} = 22.22\ \text{ft}^3 \text{ of } O_2 \quad \text{[incoming]}$$

(3.78 is the volumetric ratio of nitrogen to oxygen.)

However, some of this oxygen combines with the fuel.

$$\begin{aligned} C + O_2 &\rightarrow CO_2 \\ 2C + O_2 &\rightarrow 2CO \end{aligned}$$

1 ft^3 of O_2 can produce 1 ft^3 of CO_2 or 2 ft^3 of CO.

The oxygen volume accounted for in the stack gas is

$$\frac{9.9}{1} + 6.1 + \frac{0}{2} = 16.0\ \text{ft}^3$$

The oxygen discrepancy is

$$22.22 - 16.0 = 6.22\ \text{ft}^3 \text{ of } O_2$$

This oxygen is assumed to be combined with the hydrogen to form water vapor. By proportion, the volume of water vapor can be found.

$$\begin{aligned} 2H_2 \quad &+ \quad O_2 \quad \rightarrow \quad 2H_2O \\ 12.44\,H_2 \quad &+ \quad 6.22\,O_2 \quad \rightarrow \quad 12.44\,H_2O \end{aligned}$$

If a small error is acceptable, it can be assumed that there is no water vapor in the air. Actually, there is some. There are several ways of getting the atmospheric water.

1.0 lbm of dry air at 80°F and 14.7 psia has a volume of

$$v = \frac{RT}{p} = \frac{(53.3)(460 + 80)}{(14.7)(144)} = 13.6\ \text{ft}^3$$

From a psychrometric chart for $T_{db} = 80°F$ and $T_{wb} = 70°F$, the actual volume moist air is

$$v \approx 13.9\ \text{ft}^3$$

The volume of the water vapor is

$$v_{\text{water}} = 13.9 - 13.6 = 0.3\ \text{ft}^3/\text{lbm dry air}$$

The volume of water vapor coming in per ft^3 of dry air (the Orsat analysis basis) is

$$\frac{0.3}{13.6} = 0.022\ \text{ft}^3 \text{ of water vapor/ft}^3 \text{ dry air}$$

There are other ways of getting this quantity, including working with ω (specific humidity) and the psychrometric chart.

The atmospheric water vapor appears in the flue gas. The total water volume in the flue gas is

$$12.44 + (0.022)(22.22 + 84.0) = 14.78\ \text{ft}^3$$

The wet volumetric fraction of water vapor is

$$\frac{14.78}{100 + 14.78} = 0.129$$

The partial pressure of the water vapor is

$$(0.129)(14.5) = 1.87\ \text{psia}$$

From the steam tables, $T_{\text{saturated}}$ for 1.87 psia is 122.9°F.

The minimum stack temperature is

$$122.9 + 250 = \boxed{372.9°F}$$

2. (a) Work with 100 ft^3 of flue gas.

Nitrogen is the key element. It indicates how much air is being drawn in.

The original oxygen volume is

$$\frac{79.8}{3.78} = 21.11\ \text{ft}^3 \text{ of } O_2$$

Carbon burns according to the following.

$$\begin{aligned} C + O_2 &\rightarrow CO_2 \\ 2C + O_2 &\rightarrow 2CO \end{aligned}$$

1 ft^3 of O_2 can produce 1 ft^3 of CO_2 or 2 ft^3 of CO.

The oxygen accounted for in the Orsat analysis is

$$(1)(11.1) + \left(\tfrac{1}{2}\right)(5.3) + 0.4 = 14.15\ \text{ft}^3$$

The remaining oxygen is assumed to be in the form of water.

$$21.11 - 14.15 = 6.96\ \text{ft}^3$$

By proportions, the volume of water vapor is

$$\begin{aligned} 2H_2 \quad &+ \quad O_2 \quad \rightarrow \quad 2H_2O \\ 13.92\,H_2 \quad &+ \quad 6.96\,O_2 \quad \rightarrow \quad 13.92\,H_2O \end{aligned}$$

The wet volumetric fraction of water vapor is

$$\frac{13.92}{100 + 13.92} = 0.122$$

Since the partial pressure is weighted by mole fraction (same as volumetric fraction), the vapor pressure is

$$p_{\text{water}} = (0.122)(15) = 1.83 \text{ psia}$$

From the steam tables,

$$T_{\text{saturated}} = \boxed{121.9^\circ\text{F}}$$

(b) To produce 100 ft^3 of dry stack gases requires

$$79.8 + 21.11 = 100.91 \text{ ft}^3 \text{ of air}$$

The mass of this air is

$$m_{\text{air}} = \frac{pV}{RT} = \frac{(15)(144)(100.91)}{(53.3)(460+120)} = 7.05 \text{ lbm}$$

The fuel supplies all of the hydrogen and carbon burned.

To get 100 ft^3 of dry stack gas at 15 psia, combine 11.1 ft^3 of CO_2 at 15 psia, 5.3 ft^3 of CO at 15 psia, etc.

The masses of the gases making up the stack gas are calculated as follows.

CO_2:

$$m_{CO_2} = \frac{pV}{RT} = \frac{(15)(144)(11.1)}{(35.1)(580)} = 1.178 \text{ lbm}$$

But CO_2 is 12/(12+16+16) = 0.273 carbon by weight. The mass of carbon in the CO_2 is

$$m_C = (0.273)(1.178) = 0.3216 \text{ lbm}$$

CO:

$$m_{CO} = \frac{(15)(144)(5.3)}{(55.2)(580)} = 0.3576$$

$$m_C = \left(\frac{12}{12+16}\right)(0.3576) = 0.1533$$

CH_4:

$$m_{CH_4} = \frac{(15)(144)(1.3)}{(96.4)(580)} = 0.0502$$

CH_4 is all fuel.

Unburned H_2:

$$m_{H_2} = \frac{(15)(144)(2.1)}{(766.8)(580)} = 0.0102$$

H_2 is all fuel.

Also, there is hydrogen combined with oxygen in the form of water vapor.

$$m_{H_2O} = \frac{(15)(144)(13.92)}{(85.8)(580)} = 0.6042$$

$$m_{H_2} = \left(\frac{2}{1+1+16}\right)(0.6042) = 0.0671$$

The air/fuel ratio (by weight) is

$$R_{A/F} = \frac{7.05}{0.3216 + 0.1533 + 0.0502 + 0.0102 + 0.0671} = \boxed{11.70 \text{ lbm air/lbm fuel}}$$

3. The combustion of dodecane proceeds according to the following equation.

$$2\,C_{12}H_{26} + 37\,O_2 + 139.9\,N_2 \rightarrow 24\,CO_2 + 26\,H_2O + 139.9\,N_2$$

(2)(170)	+ (37)(32)	+ (139.9)(28)	→ (24)(44)	+ (26)(18)	+ (139.9)(28)
340	1184	3917	1056	468	3917
1	3.48	11.52	3.11	1.38	11.52

1 lbm of fuel requires stoichiometric air in the amount of

$$3.48 + 11.52 = 15.0 \text{ lbm air/lbm fuel}$$

At full load,

$$\begin{aligned}\%\text{ excess air} &= 700\% - (6)(100\%)\\ &= 100\% \quad \text{[twice the stoichiometric air]}\end{aligned}$$

The mass of air at full load is

$$\dot{m}_{\text{air}} = \frac{(2)(15)(0.43)(625)}{60\ \frac{\text{min}}{\text{hr}}} = 134.4 \text{ lbm air/min}$$

The volume of air is

$$V_{\text{air}} = \frac{mRT}{p} = \frac{(134.4)(53.3)(460+90)}{(14.7)(144)} = 1861 \text{ ft}^3/\text{min}$$

The required filter area is

$$A = \frac{V}{\text{v}} = \frac{1861\ \frac{\text{ft}^3}{\text{min}}}{600\ \frac{\text{ft}}{\text{min}}} = \boxed{3.1 \text{ ft}^2}$$

A larger filter area may be required at lower throttle settings.

4. The first step is to convert the molar fractions to mass fractions. (Molar fractions are volumetric fractions.) Take 1 mole of the mixture.

component	no. moles	MW	mass	G
CH_4	0.87	16	13.92	0.765
C_2H_6	0.04	30	1.2	0.066
C_3H_8	0.02	44	0.88	0.048
C_4H_{10}	0.005	58	0.29	0.016
C_5H_{12}	0.002	72	0.14	0.008
N_2	0.063	28	1.76	0.097
		totals	18.19	1.000

(a) Write the combustion equations and solve as a weight/proportion problem.

CH_4:

$$\begin{array}{lllll} CH_4 & + 2O_2 & \rightarrow & CO_2 + & 2H_2O \\ 16 & + 64 & \rightarrow & 44 + & 36 \end{array}$$

The oxygen required per pound of methane is

$$\frac{64}{16} = 4 \text{ lbm/lbm}$$

Air is 23.15% oxygen by weight. The air required per pound of methane is

$$\frac{4}{0.2315} = 17.28 \text{ lbm air/lbm methane}$$

The air required per pound of fuel is

$$m_{\text{air}} = (0.765)(17.28) = 13.22 \text{ lbm air/lbm fuel}$$

C_2H_6:

$$\begin{array}{lllll} 2C_2H_6 & + 7O_2 & \rightarrow & 4CO_2 + & 6H_2O \\ 60 & + 224 & \rightarrow & 176 + & 108 \end{array}$$

$$m_{\text{air}} = \frac{(0.066)\left(\frac{224}{60}\right)}{0.2315} = 1.06 \text{ lbm air/lbm fuel}$$

C_3H_8:

$$\begin{array}{lllll} C_3H_8 & + 5O_2 & \rightarrow & 3CO_2 + & 4H_2O \\ 44 & + 160 & \rightarrow & 132 + & 72 \end{array}$$

$$m_{\text{air}} = \frac{(0.048)\left(\frac{160}{44}\right)}{0.2315} = 0.754$$

C_4H_{10}:

$$\begin{array}{lllll} 2C_4H_{10} & + 13O_2 & \rightarrow & 8CO_2 + & 10H_2O \\ 116 & + 416 & \rightarrow & 352 + & 180 \end{array}$$

$$m_{\text{air}} = \frac{(0.016)\left(\frac{416}{116}\right)}{0.2315} = 0.248$$

C_5H_{12}:

$$\begin{array}{lllll} C_5H_{12} & + 8O_2 & \rightarrow & 5CO_2 + & 6H_2O \\ 72 & + 256 & \rightarrow & 220 + & 108 \end{array}$$

$$m_{\text{air}} = \frac{(0.008)\left(\frac{256}{72}\right)}{0.2315} = 0.123$$

The total air required per pound of fuel is

$$R_{A/F} = 13.22 + 1.06 + 0.754 + 0.248 + 0.123$$

$$= \boxed{15.41 \text{ lbm air/lbm fuel}}$$

(b) The stack gases will consist of N_2, CO_2, and H_2O. A stack gas analysis is not available, so use the reaction equations.

The nitrogen volume is calculated as 3.78 times the oxygen volume.

N_2:

$$\frac{\text{volume } N_2}{\text{volume fuel}} = 0.063 + (3.78)\Big[(0.87)(2) + (0.04)\left(\frac{7}{2}\right) + (0.02)(5) + (0.005)\left(\frac{13}{2}\right) + (0.002)(8)\Big]$$

$$= 7.731 \text{ ft}^3 \text{ } N_2/\text{ft}^3 \text{ fuel} \quad [\text{or moles } N_2/\text{mole fuel}]$$

CO_2:

$$\frac{\text{volume } CO_2}{\text{volume fuel}} = (0.87)(1) + (0.04)\left(\frac{4}{2}\right) + (0.02)\left(\frac{3}{1}\right) + (0.005)\left(\frac{8}{2}\right) + (0.002)\left(\frac{5}{1}\right)$$

$$= 1.04 \text{ ft}^3 \text{ } CO_2/\text{ft}^3 \text{ fuel}$$

H_2O:

$$\frac{\text{volume } H_2O}{\text{volume fuel}} = (0.87)\left(\frac{2}{1}\right) + (0.04)\left(\frac{6}{2}\right) + (0.02)\left(\frac{4}{1}\right) + (0.005)\left(\frac{10}{2}\right) + (0.002)\left(\frac{6}{1}\right)$$

$$= 1.98 \text{ ft}^3 \text{ } H_2O/\text{ft}^3 \text{ fuel}$$

The partial pressure is proportional to the mole (volume) fraction.

$$x_{H_2O} = \frac{1.98}{7.73 + 1.04 + 1.98} = 0.184$$

$$p_{H_2O} = (0.184)(14.4) = 2.65 \text{ psia}$$

From the steam tables,

$$T_{\text{saturated}}\Big|_{2.65 \text{ psia}} = \boxed{136°\text{F}}$$

(c) Excess air will lower x_{H_2O} and p_{H_2O}, lowering the dew point. The presence of atmospheric moisture will also lower it.

To avoid corrosion, keep the temperature above 136°F.

Other considerations are equally important, for example, the temperature that produces an adequate draft. The dew point may not be the determining factor.

5. (a) The combustion reaction is

$$\underset{224}{C_{16}H_{32}}+\underset{768}{24\,O_2}+\underset{2540}{(3.78)(24)N_2} \rightarrow \underset{704}{16\,CO_2}+\underset{288}{16\,H_2O}+\underset{2540}{(3.78)(24)N_2}$$

The air required per pound of fuel is

$$\frac{768+2540}{224} = 14.77 \text{ lbm air/lbm fuel}$$

The mass of air required is (40)(14.77 lbm/min). The corresponding volume is

$$V_{\text{air}} = \frac{mRT}{p} = \frac{(40)(14.77)(53.3)(460+60)}{(14.7)(144)} = \boxed{7736 \text{ ft}^3/\text{min}}$$

(b) Assuming no energy losses, the energy released in combustion is

$$\left(\frac{40}{8.23}\right)(152{,}000) = 7.39 \times 10^5 \text{ BTU/min}$$

The air and combustion gas temperature increases from 60°F to 1350°F.

$$c_p = \frac{Rk}{J(k-1)} = \frac{\left(\frac{1545}{32.4}\right)(1.4)}{(778)(1.4-1)} = 0.2145 \text{ BTU/lbm-°R}$$

From $q = mc_p\Delta T$,

$$m_{\text{combustion gases}} = \frac{q}{c_p\Delta T} = \frac{7.39 \times 10^5}{(0.2145)(1350-60)} = 2671 \text{ lbm/min}$$

Per pound of fuel, the mass of combustion gases is

$$\frac{2671}{40} = \boxed{66.8 \text{ lbm gases/min-lbm fuel}}$$

Notice that the amount of excess air is unknown, so the combustion reaction equation could not be used.

6. Assume the following.

- Mass of fuel is small compared to mass of air.
- Combustion is a constant pressure process.
- Combustion is adiabatic.
- Combustion is complete.
- There is no nitrogen dissociation.
- Air is dry.
- Gases are ideal (constant c_p).
- Heat of vaporization is small for fuel.

First iteration:

The composition of the stack gases (including the percentage of excess oxygen) is unknown, so c_p of stack gases cannot be found.

Assume combustion products have the same specific heat as nitrogen. (Stack gases are primarily nitrogen.)

The average temperature of the nitrogen is

$$\left(\tfrac{1}{2}\right)(2250+575) = 1412.5\text{°F} = 1873\text{°R}$$

At 1873°R,

$$c_{p,N_2} \approx 0.280 \text{ BTU/lbm-°F}$$

The combustion heat required is

$$q = \dot{m}c_p\Delta T = (11.4)(0.280)(2250-575) = 5347 \text{ BTU/sec}$$

Use the heating value of iso-octane.

$$\dot{m}_{\text{fuel}} = \frac{q}{\text{LHV}} = \frac{5347}{19{,}160} = 0.279 \text{ lbm/sec}$$

(Note that LHV should be used.)

The air/fuel ratio is

$$R_{A/F} = \frac{11.4}{0.279} = \boxed{40.86 \text{ lbm air/lbm fuel}}$$

Second iteration:

- Use the actual stack gases.
- Include the fuel mass.

Calculate the amount of excess air.

$$\underset{228}{2\,C_8H_{18}} + \underset{800}{25\,O_2} \rightarrow \underset{704}{16\,CO_2} + \underset{324}{18\,H_2O}$$

The stoichiometric air/fuel ratio is

$$R_{A/F,\text{ideal}} = \frac{800}{(0.2315)(228)} = 15.2 \text{ lbm air/lbm fuel}$$

The ratio of supplied air to ideal air is

$$\frac{40.86}{15.2} = 2.69 \quad [169\% \text{ excess air}]$$

Note that this could not be calculated without doing the first iteration. The complete combustion equation is

$$\underset{228}{2\,C_8H_{18}} + \underset{2152}{(2.69)(25)O_2} + \underset{7118}{(2.69)(3.78)(25)N_2}$$

$$\rightarrow \underset{704}{16\,CO_2} + \underset{324}{18\,H_2O} \; \underset{1352}{(1.69)(25)O_2} + \underset{7118}{(2.69)(3.78)(25)N_2}$$

The gravimetric fractions of the combustion products are

CO_2:

$$\frac{704}{704+324+1352+7118} = \frac{704}{9498} = 0.074$$

H_2O:

$$\frac{324}{9498} = 0.034$$

O_2:

$$\frac{1352}{9498} = 0.142$$

N_2:

$$\frac{7118}{9498} = 0.749$$

At 1873°R, the approximate specific heats are

CO_2: 0.297
H_2O: 0.554
O_2: 0.261
N_2: 0.280

The average specific heat of the combustion gases is

$$\begin{aligned}\overline{c_p} &= (0.074)(0.297) + (0.034)(0.554) \\ &\quad + (0.142)(0.261) + (0.749)(0.280) \\ &= 0.288\end{aligned}$$

The approximate fuel mass (from iteration 1) is 0.279 lbm/sec.

The required heat energy is

$$\begin{aligned}q = mc_p\Delta T &= (11.4 + 0.279)(0.288)(2250 - 575) \\ &= 5634 \text{ BTU/sec}\end{aligned}$$

$$m_{\text{fuel}} = \frac{5634}{19{,}160} = 0.294 \text{ lbm/sec}$$

The adjusted air/fuel ratio is

$$R_{\text{A/F}} = \frac{11.4}{0.294} = \boxed{38.78 \text{ lbm air/lbm fuel}}$$

Another iteration would refine this further.

7. Take 1 lbm of corn cobs.

The stoichiometric air is

$$(34.34)\left[\frac{0.48}{3} + \left(0.05 - \frac{0.40}{8}\right)\right] = 5.49 \text{ lbm air/lbm fuel}$$

Note that all of the oxygen and hydrogen are locked up in the form of water.

When 1 lbm of corn cobs burns, the energy goes into

- heating incoming air
- vaporizing liquid water (0.05 + 0.40 lbm of water)
- heating the building (to be neglected)

(a) The composition of the stack gases is unknown, so c_p of stack gases cannot be found.

Assume stack gas and ash have the same specific heat as nitrogen.

The average temperature is

$$\left(\tfrac{1}{2}\right)(34 + 450) = 242°\text{F} = 702°\text{R}$$

At 702°R,

$$c_{p,N_2} = 0.251 \text{ BTU/lbm-°R}$$

The heat required to bring the air, corn, and moisture combustion products up to 450°F is

$$\begin{aligned}q = mc_p\Delta T &= (5.49 + 1 - 0.45)(0.251)(450 - 34) \\ &= 631 \text{ BTU/lbm corn}\end{aligned}$$

The water in this corn has an initial enthalpy of about 2 BTU/lbm (i.e., 34°F − 32°F = 2°F).

At 450°F and low pressure, the steam enthalpy is approximately

$$h \approx 1264 \text{ BTU/lbm} \quad \text{[not highly pressure sensitive]}$$

The energy required to vaporize the water is

$$(0.45)(1264 - 2) = 568 \text{ BTU/lbm corn}$$

The net energy available to heat the building is

$$6600 - 631 - 568 = 5401 \text{ BTU/lbm corn}$$

If this energy is not absorbed by the building, it will have to be absorbed by excess air. Assume $c_{p,\text{air}} = c_{p,N_2}$.

$$\begin{aligned}m_{\text{air, excess}} = \frac{q}{c_p\Delta T} &= \frac{5401}{(0.251)(450 - 34)} \\ &= 51.7 \text{ lbm air/lbm corn}\end{aligned}$$

The excess air is

$$\% \text{ excess air} = \left(\frac{51.7}{5.49}\right)(100\%)$$

$$= \boxed{942\%}$$

Knowing the percent of excess air, c_p of the stack gases could be revised.

(b) $m_{\text{air}} = (500)(5.49 + 51.7) = 28{,}595 \text{ lbm/air}$

$$V = \frac{mRT}{p} = \frac{(28{,}595)(53.3)(460 + 34)}{(14.7)(144)}$$

$$= \boxed{3.56 \times 10^5 \text{ ft}^3}$$

HEAT TRANSFER

1. Check the Biot modulus. The characteristic length is

$$L = \frac{V}{A_s} = \frac{\frac{4}{3}\pi r^3}{4\pi r^2} = \frac{r}{3}$$

$$= \frac{\frac{3.5}{12}}{(2)(3)} = 0.0486 \text{ ft}$$

$$\text{Bi} = \frac{hL}{k} = \frac{(5.7)(0.0486)}{0.4} = 0.693$$

Since Bi > 0.1, charts must be used for the solution.

$$\frac{T_t - T_\infty}{T_0 - T_\infty} = \frac{40-10}{80-10} = 0.43$$

$$r_o = \frac{\frac{3.5}{12}}{2} = 0.146 \text{ ft}$$

$$\frac{k}{hr_o} = \frac{0.42}{(5.7)(0.146)} = 0.50$$

From a transient chart Fo ≈ 0.30 (relative time).

From the definition of Fourier number, solving t,

$$t = \frac{r_o^2 \text{Fo}}{a} = \frac{(0.146)^2(0.30)}{0.0063}$$

$$= \boxed{1.02 \text{ hr}}$$

2. Heat is lost from the top and sides.

$$q = 5 \text{ W} = (5 \text{ W})\left(3.412 \, \frac{\text{BTU}}{\text{hr-W}}\right) = 17.06 \text{ BTU/hr}$$

$$A_{\text{end}} = \left(\frac{\pi}{4}\right)\left(\frac{0.80}{12}\right)^2 = 3.49 \times 10^{-3} \text{ ft}^2$$

$$A_{\text{side}} = \pi\left(\frac{0.80}{12}\right)\left(\frac{1.7}{12}\right) = 0.0297 \text{ ft}^2$$

$$q = A_{\text{end}}h_{\text{end}}(T_s - T_\infty) + A_{\text{side}}h_{\text{side}}(T_s - T_\infty) + \sigma F_e F_A (A_{\text{end}} + A_{\text{side}})(T_s^4 - T_\infty^4)$$

h is unknown for both the end and side. It cannot be found until T_s is known.

Initial guess: $h = 1.65$ BTU/hr-ft^2-°F.

$$T_\infty = 120 + 460 = 580°\text{R}$$

$$F_e = \epsilon_1 = 0.65$$

$$F_A = 1 \quad \text{[enclosure]}$$

$$17.06 = (3.49 \times 10^{-3})(1.65)(T_s - 580) + (0.0297)(1.65)(T_s - 580) + (0.65)(1)(0.0332)(0.1713 \times 10^{-8}) \times [T_s^4 - (580)^4]$$

By trial and error,

$$T_s \approx 750°\text{R}$$

If $T_s = 750°$R, then

$$T_{\text{film}} = \left(\tfrac{1}{2}\right)(750 + 580) = 665°\text{R} = 205°\text{F}$$

At 205°F, for air,

$$\text{Pr} = 0.72$$

$$\frac{g\beta\rho^2}{\mu^2} = 0.848 \times 10^6$$

The characteristic lengths are

$$L_{\text{end}} = \text{diameter} = \frac{0.80}{12} = 0.0667 \text{ ft}$$

$$L_{\text{side}} = \text{height} = \frac{1.7}{12} = 0.142 \text{ ft}$$

The Grashof numbers are

$$\text{Gr} = \frac{L^3\rho^2\beta\Delta T g}{\mu^2}$$

$$\text{Gr}_{\text{end}} = (0.0667)^3(0.848 \times 10^6)(750 - 580) = 4.28 \times 10^4$$

$$\text{Gr}_{\text{sides}} = (0.142)^3(0.848 \times 10^6)(750 - 580) = 4.13 \times 10^5$$

$$(\text{GrPr})_{\text{end}} = (4.28 \times 10^4)(0.72) = 3.08 \times 10^4$$

$$(\text{GrPr})_{\text{sides}} = (4.13 \times 10^5)(0.72) = 2.97 \times 10^5$$

Using a correlation for air,

$$h_{\text{end}} = (0.27)\left(\frac{750 - 580}{0.0667}\right)^{0.25} = 1.92$$

$$h_{\text{side}} = (0.29)\left(\frac{750 - 580}{0.142}\right)^{0.25} = 1.71$$

$$17.06 = (3.49 \times 10^{-3})(1.92)(T_s - 580) + (0.0297)(1.71)(T_s - 580) + (0.65)(0.0332)(0.1713 \times 10^{-8}) \times [T_s^4 - (580)^4]$$

By trial and error,

$$T_s \approx \boxed{748°\text{R} = 288°\text{F}}$$

(b)
$$q_{\text{convection}} = q_{\text{total}} - q_{\text{radiation}} = 17.06 - (0.65)(0.0332) \times (0.1713 \times 10^{-8})[(748)^4 - (580)^4] = 9.67$$

$$\% \text{ convection} = \frac{9.67}{17.06} = 0.567 \quad \boxed{56.7\%}$$

3.

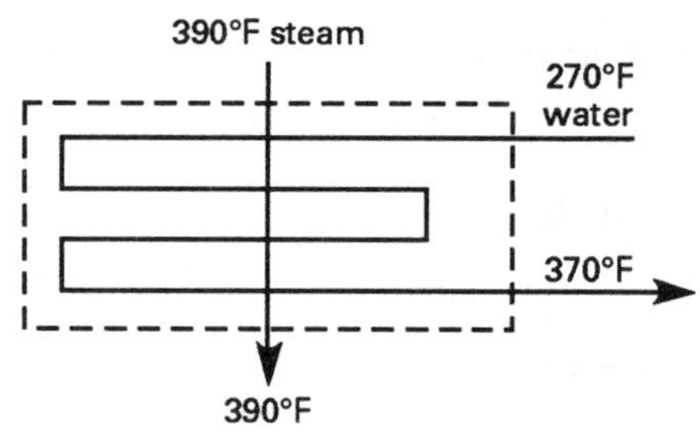

(a) For the water, use $q = \Delta h$.

$$h_{\text{in},270°\text{F}} = 238.84 \text{ BTU/lbm}$$
$$h_{\text{out},370°\text{F}} = 342.79 \text{ BTU/lbm}$$
$$v_{\text{F,out}} = 0.01823 \text{ ft}^3\text{/lbm} \quad [\text{larger than } v_{\text{in}}]$$
$$q = \Delta h = 342.79 - 238.84 = 103.95 \text{ BTU/lbm}$$
$$\dot{q} = (400{,}000)(103.95) = 4.16 \times 10^7 \text{ BTU/hr}$$

The required flow area to keep the velocity at 5 ft/sec is

$$A = \frac{Q}{\text{v}} = \frac{(400{,}000)(0.01823)}{(5)(3600)} = 0.405 \text{ ft}^2$$

The flow area per tube is

$$A_{\text{tube}} = \left(\frac{\pi}{4}\right)\left[\frac{1-(2)\left(\frac{1}{16}\right)}{12}\right]^2 = 0.00418 \text{ ft}^2\text{/tube}$$

The number of tubes is

$$\frac{0.405}{0.00418} = \boxed{96.9 \text{ (say 100) per pass}}$$

Actually, 20–30% more could be added for future plugging.

(b) The number of tubes is

$$(4)(100) = \boxed{400 \text{ tubes}}$$

(c) Since U is given, h_i and h_o do not need to be calculated.

$$\Delta T_{\text{in}} = 390 - 270 = 120°$$
$$\Delta T_{\text{out}} = 390 - 370 = 20°$$
$$\Delta T_m = \frac{120-20}{\ln\left(\frac{120}{20}\right)} = \frac{100}{1.79} = 55.8°$$
$$F_c = 1 \text{ for condensing steam}$$
$$q = UAF_c\Delta T_m$$
$$4.16 \times 10^7 = (700)A(1)(55.8)$$
$$A = 1065 \text{ ft}^2 \quad [\text{total}]$$

The area per tube is

$$\frac{1065}{400} = 2.66 \text{ ft}^2$$

The outside surface area per foot of tube length is

$$A = \pi DL = \pi\left(\frac{1}{12}\right)(1) = 0.2618 \text{ ft}^2$$
$$L_{\text{tube}} = \frac{2.66}{0.2618} = 10.16 \text{ ft}$$

The exchanger length is

$$L_{\text{exchanger}} = 3 + 10.16 = \boxed{13.16 \text{ ft}}$$

4.
$$r_{\text{critical}} = \frac{k}{h} = \frac{0.03}{2.0} = 0.015 \text{ ft}$$

The wire radius is $0.012/2 = 0.006$.

The insulation thickness is

$$0.015 - 0.006 = \boxed{0.009 \text{ ft}}$$

5. This is a combined heat transfer problem, as radiation and convection both contribute to the heat dissipation. Normally, radiation would not be considered at the low wall temperatures expected. However, the overall heat transfer is itself low, so radiation might be significant.

The heat transfer quantities from these two mechanisms are highly sensitive to the surface temperature. Though an iterative procedure would eventually work, it is more expedient to consider the two mechanisms separately.

Assume $T_s = 80°\text{F}$.

Calculate the radiation heat transfer per unit area.

$$\begin{aligned} q_{\text{radiation}} &= \sigma F_e F_a (T_s^4 - T_\infty^4) \\ &= \left(0.1713 \times 10^{-8} \ \frac{\text{BTU}}{\text{hr-ft}^2\text{-°R}^4}\right) \\ &\quad \times (0.90)(1.00)(1 \text{ ft}^2) \\ &\quad \times \left[(80°\text{F} + 460)^4 - (70°\text{F} + 460)^4\right] \\ &= 9.44 \text{ BTU/hr} \end{aligned}$$

Calculate the convection heat transfer per unit area.

$$\begin{aligned}
T_{\text{film}} &= \left(\tfrac{1}{2}\right)(T_s + T_\infty) \\
&= \left(\tfrac{1}{2}\right)(80^\circ\text{F} + 70^\circ\text{F}) \\
&= 75^\circ\text{F} \\
\text{Pr} &= 0.72 \\
\frac{g\beta\rho^2}{\mu^2} &= 2.27 \times 10^6 \ 1/\text{ft}^3\text{-}^\circ\text{F} \quad \text{[interpolated]} \\
\Delta T &= T_s - T_\infty \\
&= 80^\circ\text{F} - 70^\circ\text{F} \\
&= 10^\circ\text{F} \\
L &= 5 \text{ ft} \quad \text{[vertical height]} \\
\text{Gr} &= (L^3)\left(\frac{g\beta\rho^2}{\mu^2}\right)\Delta T \\
&= (5 \text{ ft})^3 \left(2.27 \times 10^6 \ \frac{1}{\text{ft}^3\text{-}^\circ\text{F}}\right)(10\ ^\circ\text{F}) \\
&= 2.84 \times 10^9 \\
\text{PrGr} &= (0.72)(2.84 \times 10^9) \\
&= 2.04 \times 10^9 \\
h &\approx (0.19)(\Delta T)^{0.33} \\
&= (0.19)(10^\circ\text{F})^{0.33} \\
&= 0.406 \text{ BTU/hr-ft}^2\text{-}^\circ\text{F} \\
q_{\text{convection}} &= hA\Delta T \\
&= \left(0.406 \ \frac{\text{BTU}}{\text{hr-ft}^2\text{-}^\circ\text{F}}\right)(1 \text{ ft}^2)(10^\circ\text{F}) \\
&= 4.06 \text{ BTU/hr} \\
q_{\text{total},80^\circ\text{F}} &= q_{\text{radiation}} + q_{\text{convection}} \\
&= 9.44 \ \frac{\text{BTU}}{\text{hr}} + 4.06 \ \frac{\text{BTU}}{\text{hr}} \\
&= 13.5 \text{ BTU/hr}
\end{aligned}$$

Assume $T_s = 90^\circ\text{F}$.

Calculate the radiation heat transfer per unit area.

$$\begin{aligned}
q_{\text{radiation}} &= \sigma F_e F_a (T_s^4 - T_\infty^4) \\
&= \left(0.1713 \times 10^{-8} \ \frac{\text{BTU}}{\text{hr-ft}^2\text{-}^\circ\text{R}^4}\right) \\
&\quad \times (0.90)(1.00)(1 \text{ ft}^2) \\
&\quad \times \left[(90\ ^\circ\text{F} + 460)^4\right. \\
&\quad \left. -(70^\circ\text{F} + 460)^4\right] \\
&= 19.43 \text{ BTU/hr}
\end{aligned}$$

Calculate the convective heat transfer per unit area.

$$\begin{aligned}
T_{\text{film}} &= \left(\tfrac{1}{2}\right)(90^\circ\text{F} + 70^\circ\text{F}) \\
&= 80^\circ\text{F} \\
\text{Pr} &= 0.72 \\
\frac{g\beta\rho^2}{\mu^2} &= 2.17 \times 10^6 \ 1/\text{ft}^3\text{-}^\circ\text{F} \\
\Delta T &= 90^\circ\text{F} - 70^\circ\text{F} \\
&= 20^\circ\text{F} \\
L &= 5 \text{ ft} \\
\text{Gr} &= (5 \text{ ft})^3 \left(2.17 \times 10^6 \ \frac{1}{\text{ft}^3\text{-}^\circ\text{F}}\right)(20^\circ\text{F}) \\
&= 5.43 \times 10^9 \\
\text{PrGr} &= (0.72)(5.43 \times 10^9) \\
&= 3.9 \times 10^9 \\
h &= (0.19)(20^\circ\text{F})^{0.33} \\
&= 0.511 \text{ BTU/hr-ft}^2\text{-}^\circ\text{F} \\
q_{\text{convective}} &= \left(0.511 \ \frac{\text{BTU}}{\text{hr-ft}^2\text{-}^\circ\text{F}}\right)(1 \text{ ft}^2)(20^\circ\text{F}) \\
&= 10.22 \text{ BTU/hr} \\
q_{\text{total},90\ ^\circ\text{F}} &= 19.43 \ \frac{\text{BTU}}{\text{hr}} + 10.22 \ \frac{\text{BTU}}{\text{hr}} \\
&= 29.65 \text{ BTU/hr}
\end{aligned}$$

The actual heat dissipated per square foot is

$$q_{\text{actual}} = \frac{(200 \text{ W})\left(3.413 \ \frac{\text{BTU}}{\text{hr-W}}\right)}{(2 \text{ ft})(5 \text{ ft})(2 \text{ sides})} = 34.13 \text{ BTU/hr}$$

(c) Extrapolating to obtain the surface temperature that would result in this value,

$$\begin{aligned}
T_s &= 90^\circ\text{F} + (10^\circ\text{F})\left(\frac{34.13 \ \frac{\text{BTU}}{\text{hr}} - 29.65 \ \frac{\text{BTU}}{\text{hr}}}{29.65 \ \frac{\text{BTU}}{\text{hr}} - 13.5 \ \frac{\text{BTU}}{\text{hr}}}\right) \\
&= \boxed{92.8^\circ\text{F} \quad \text{[use 93°F]}}
\end{aligned}$$

(a) Now, work with the sheetrock only. Notice that the value of k requires L to be in units of inches.

$$\begin{aligned}
L &= 1 \text{ in} \\
\Delta T &= T_{\text{inside}} - T_s = T_{\text{inside}} - 93^\circ\text{F} \\
q &= \frac{kA\Delta T}{L} \\
34.13 \ \frac{\text{BTU}}{\text{hr}} &= \frac{\left(1.0 \ \frac{\text{BTU-in}}{\text{hr-ft}^2}\right)(1 \text{ ft}^2)(T_{\text{inside}} - 93^\circ\text{F})}{1 \text{ in}} \\
T_{\text{inside}} &= \boxed{127.13^\circ\text{F} \quad \left[\begin{array}{c}\text{at interface of} \\ \text{steel and sheetrock}\end{array}\right]}
\end{aligned}$$

(b) This is a case of internal heat generation. The energy generated per cubic foot of steel is

$$q^* = \frac{q}{V}$$

$$= \frac{(200 \text{ W})\left(3.413 \ \frac{\text{BTU}}{\text{hr-W}}\right)}{(2 \text{ ft})(5 \text{ ft})\left(\frac{\frac{1}{4} \text{ in}}{12 \ \frac{\text{in}}{\text{ft}}}\right)}$$

$$= 3276.5 \text{ BTU/hr-ft}^3$$

The distance from the center of the plate to the nearest surface is

$$L = \frac{t}{2}$$

$$= \frac{\frac{1}{4} \text{ in}}{(2)\left(12 \ \frac{\text{in}}{\text{ft}}\right)} = 1.0417 \times 10^{-2} \text{ ft}$$

$$k_{\text{steel}} \approx 26 \text{ BTU/hr-ft-°F}$$

$$T_{\text{center}} = T_s + \frac{q^* L^2}{2k}$$

$$= 127.3\text{°F} + \frac{\left(3276.5 \ \frac{\text{BTU}}{\text{hr-ft}^3}\right)(1.0417 \times 10^{-2} \text{ ft})^2}{(2)\left(26 \ \frac{\text{BTU}}{\text{hr-ft-°F}}\right)}$$

$$= \boxed{127.3\text{°F} \quad \left[\begin{array}{c}\text{essentially uniform} \\ \text{temperature throughout}\end{array}\right]}$$

6. As tested:

$$\Delta T_A = 240 - 70 = 170\text{°F}$$

$$\Delta T_B = 240 - 150 = 90\text{°F}$$

$$\Delta T_M = \frac{170 - 90}{\ln\left(\frac{170}{90}\right)} = 125.8\text{°F}$$

$$F_c = 1 \quad \text{[condensing steam]}$$

At 70°F,

$$v_F = 0.01606 \text{ ft}^3\text{/lbm}$$

$$\dot{m} = \frac{(120 \text{ gpm})\left(0.1337 \ \frac{\text{ft}^3}{\text{gal}}\right)\left(60 \ \frac{\text{min}}{\text{hr}}\right)}{0.01606 \ \frac{\text{ft}^3}{\text{lbm}}}$$

$$= 59{,}940 \text{ lbm/hr}$$

Notice that it is assumed that 120 gpm of 70°F water is heated. Thus, v_F at 70°F is required.

To avoid having to use an average c_p to find q, use enthalpies.

At 70°F,

$$h = 38.04 \text{ BTU/lbm}$$

At 150°F,

$$h = 117.89 \text{ BTU/lbm}$$

$$\dot{q} = \dot{m}\Delta h$$

$$= (59{,}940)(117.89 - 38.04)$$

$$= 4.79 \times 10^6 \text{ BTU/hr}$$

Then from $q = UAF_c\Delta T_m$,

$$U = \frac{4.79 \times 10^6}{(60)(1)(125.8)} = 634.6 \text{ BTU/hr-ft}^2\text{-°F}$$

After 2 years:

$$\Delta T_A = 250 - 70 = 180$$

$$\Delta T_B = 250 - 130 = 120$$

$$\Delta T_M = \frac{180 - 120}{\ln\left(\frac{180}{120}\right)} = 148.0$$

$$F_c = 1$$

At 70°F,

$$v_F = 0.01606 \text{ ft}^3\text{/lbm}$$

$$\dot{m} = \frac{(100)(0.1337)(60)}{0.01606} = 49{,}950 \text{ lbm/hr}$$

$$h_{70\text{°F}} = 38.04 \text{ BTU/lbm}$$

$$h_{130\text{°F}} = 97.90 \text{ BTU/lbm}$$

$$q = (49{,}950)(97.9 - 38.04) = 2.99 \times 10^6 \text{ BTU/hr}$$

$$U = \frac{q}{AF_c\Delta T_m} = \frac{2.99 \times 10^6}{(60)(1)(148)} = 336.7$$

$$\frac{1}{U_{\text{fouled}}} = \frac{1}{U_{\text{clean}}} + R_f$$

$$R_f = \frac{1}{336.7} - \frac{1}{634.6}$$

$$= \boxed{0.00139 \text{ ft}^2\text{-hr-°F/BTU}}$$

This assumes h_o and h_i do not change.

7.

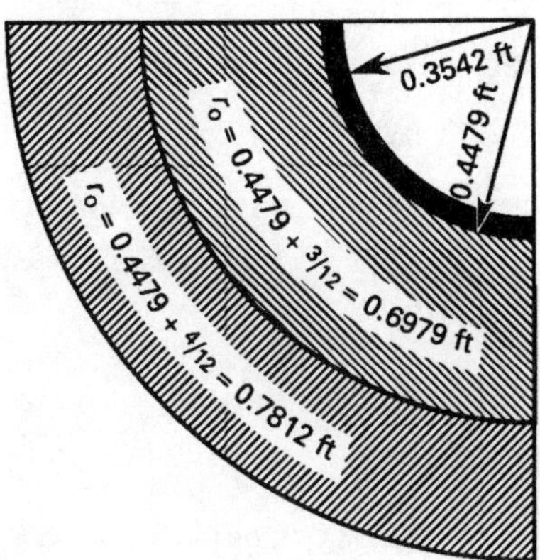

First, gather the required data. For schedule-160, 10-in pipe,

$$r_i = \frac{0.7083}{2} = 0.3542 \text{ ft}$$

$$r_o = \frac{10.75}{(12)(2)} = 0.4479 \text{ ft}$$

The approximate thermal conductivity must be used since the actual temperature is not known.

$$k_{\text{magnesia}} \approx 0.037 \text{ BTU-ft/ft}^2\text{-hr-}^\circ\text{F}$$
$$k_{\text{earth}} \approx 0.036$$

For oxidized steel tube,

$$\epsilon_{\text{steel}} \approx 0.8 \quad \text{[approximate]}$$

(a) Without insulation:

The loss at 400°F would include radiation and natural convection. Assume $T_{\text{pipe}} = 400°\text{F}$.

$$T_{\text{film}} = \left(\tfrac{1}{2}\right)(400 + 70) = 235°\text{F}$$

For air at 235°F.

$$\text{Pr} = 0.72 \quad \text{[interpolated]}$$

$$\frac{g\beta\rho^2}{\mu^2} = 0.708 \times 10^6 \quad \text{[interpolated]}$$

$$L = \text{pipe diameter} = \frac{10.75}{12} = 0.896 \text{ ft}$$

$$\begin{aligned}\text{Gr} &= (0.896)^3(0.708 \times 10^6)(400 - 70) \\ &= 1.68 \times 10^8\end{aligned}$$

$$\begin{aligned}\text{PrGr} &= (0.72)(1.68 \times 10^8) \\ &= 1.21 \times 10^8\end{aligned}$$

From a useful correlation for air,

$$h \approx (0.27)\left(\frac{400 - 70}{0.896}\right)^{0.25} = 1.18 \text{ BTU/hr-ft}^2\text{-}^\circ\text{F}$$

For 1 ft of length,

$$A_o = \pi(0.896)(1) = 2.815 \text{ ft}^2$$

$$\begin{aligned}q_{\text{total}} &= (1.18)(2.815)(400 - 70) \\ &\quad + (2.815)(0.1713 \times 10^{-8})(0.8) \\ &\quad \times \left[(400 + 460)^4 - (70 + 460)^4\right] \\ &= 1096 + 1806 \\ &= 2902 \text{ BTU/hr-ft pipe}\end{aligned}$$

With insulation:

- first approximation
- ignore radiation
- T_s is unknown, so assume $h_o = 2.0$
- disregard steel heat transfer resistance
- combine two insulations into a single layer since k values are similar

$$\begin{aligned}q_{\text{total}} &= \frac{2\pi(1)(400 - 70)}{\dfrac{\ln\left(\dfrac{0.7812}{0.4479}\right)}{0.036} + \dfrac{1}{(0.7812)(2)}} \\ &= 129 \text{ BTU/hr}\end{aligned}$$

Calculate the surface temperature.

$$129 = \frac{2\pi(1)(400 - T_s)}{\dfrac{\ln\left(\dfrac{0.7812}{0.4479}\right)}{0.036}}$$

$$T_s = 82.8 \quad \text{[say 83°F]}$$

The assumption that radiation is insignificant is valid.

The new T_s could be used to obtain a more accurate value of k. But k does not vary much, and it is unlikely the additional sophistication is warranted.

The film temperature is

$$T_{\text{film}} = \left(\tfrac{1}{2}\right)(83 + 70) = 77°\text{F}$$

At 77°F,

$$\text{Pr} = 0.72 \quad \text{[interpolated]}$$

$$\frac{g\beta\rho^2}{\mu^2} = 2.23 \times 10^6 \quad \text{[interpolated]}$$

The outside insulation diameter is

$$D = \frac{10.75 + 8}{12} = 1.56 \text{ ft}$$

$$\begin{aligned}\text{Gr} &= (1.56)^3(2.23 \times 10^6)(83 - 70) \\ &= 1.1 \times 10^8\end{aligned}$$

$$\begin{aligned}\text{PrGr} &= (0.72)(1.1 \times 10^8) \\ &= 7.9 \times 10^7\end{aligned}$$

Use a correlation for air.

$$h = (0.27)\left(\frac{83-70}{1.56}\right)^{0.25} = 0.46$$

The heat flow is

$$q_{\text{total}} = \frac{2\pi(1)(400-70)}{\frac{\ln\left(\frac{0.7812}{0.4479}\right)}{0.036} + \frac{1}{(0.7812)(0.46)}}$$
$$= 114 \text{ BTU/hr}$$

The efficiency is

$$\eta = \frac{2902-114}{2902} = \boxed{0.961}$$

(b) This is a capitalized cost problem. Assume an infinite life.

$$\text{CC} = \frac{(2902-114)(24)(365)(2.46\times10^{-6})}{0.20}$$
$$= \boxed{\$300 \text{ per foot of pipe}}$$

This is the present worth of a project with an infinite life.

8. This is a difficult problem because there are two variables: surface temperature and the heat flow split between the sides and the top.

Both must be balanced simultaneously to keep q at a known value.

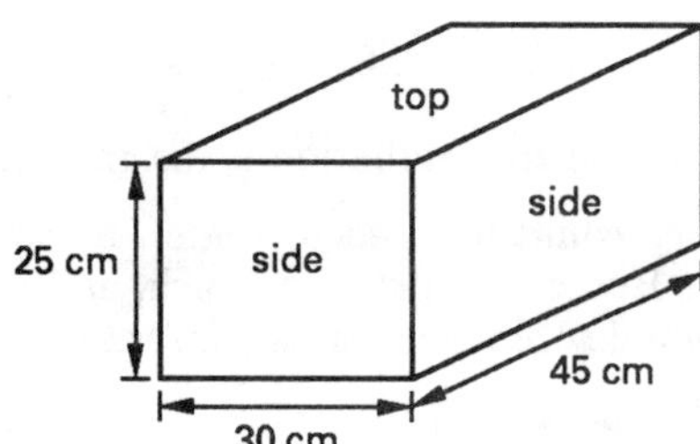

This would be trivial unless different film coefficients are considered for the top and sides.

Convert to English units.

$$q = (200)(3.413) = 682.6 \text{ BTU/hr}$$

(Multiply centimeters by 3.281×10^{-2} to get feet.)

As a first approximation, assume heat loss is proportional to area, regardless of orientation.

$$A_{\text{top}} = (30)(45)(3.281\times10^{-2})^2 = 1.45 \text{ ft}^2$$
$$A_{\text{sides}} = (30+30+45+45)(25)(3.281\times10^{-2})^2$$
$$= 4.04 \text{ ft}^2$$
$$q_{\text{top}} \approx \left(\frac{1.45}{1.45+4.04}\right)(682.6) = 180.3 \text{ BTU/hr}$$

Assume $h_{\text{top}} = 1.65$ BTU/hr-ft^2-°F.

$$T_\infty = 32 + \left(\frac{9}{5}\right)(20°\text{C}) = 68°\text{F}$$
$$q = hA(T_s - T_\infty)$$
$$180.3 = (1.65)(1.45)(T_{s,\text{top}} - 68)$$
$$T_{s,\text{top}} \approx 143°\text{F}$$
$$T_{\text{film,top}} = \left(\tfrac{1}{2}\right)(68+143) = 105.5°\text{F} \quad \text{[say 106]}$$

From air at 106°F,

$$\text{Pr} = 0.72$$
$$\frac{g\beta\rho^2}{\mu^2} = 1.71\times10^6$$
$$L = \left(\tfrac{1}{2}\right)(30+45)(3.281\times10^{-2}) = 1.23 \text{ ft}$$
$$\text{Gr} = (1.23)^3(1.71\times10^6)(143-68)$$
$$= 2.39\times10^8$$
$$\text{PrGr} = (0.72)(2.39\times10^8)$$
$$= 1.7\times10^8$$

From a correlation for air,

$$h_{\text{top}} = (0.22)(T_s - T_\infty)^{0.33}$$
$$= (0.22)(143-68)^{0.33}$$
$$= 0.91$$

Calculate a new T_s assuming the same split of heat loss.

$$q = hA(T_s - T_\infty)$$
$$180.3 = (0.91)(1.45)(T_s - 68)$$
$$T_s = 204.6 \quad \text{[say 200°F]}$$

If the surface temperature is 200°F, for the top,

$$\text{Pr} = 0.72$$
$$\frac{g\beta\rho^2}{\mu^2} = 0.85\times10^6$$
$$L_{\text{top}} = 1.23 \text{ ft}$$
$$\text{Gr} = (1.23)^3(0.85\times10^6)(200-68)$$
$$= 2.09\times10^8$$
$$\text{PrGr} = (0.72)(2.09\times10^8)$$
$$= 1.5\times10^8$$

Use a correlation for air.

$$h_{\text{top}} = (0.22)(T_s - 68)^{0.33}$$

For the sides,

$$L_{\text{sides}} = (25)(3.281\times10^{-2})$$
$$= 0.82 \text{ ft}$$
$$\text{Gr} = (0.82)^3(0.85\times10^6)(200-68)$$
$$= 6.19\times10^7$$
$$\text{PrGr} = (0.72)(6.19\times10^7)$$
$$= 4.45\times10^7$$

Use a correlation for air.

$$h_{\text{sides}} = (0.29)\left(\frac{T_s - 68}{0.82}\right)^{0.25}$$

The total heat loss is known, so

$$q = h_{\text{total}}A_{\text{top}}(T_s - 68) + h_{\text{sides}}A_{\text{sides}}(T_s - 68)$$

$$682.6 = (0.22)(T_s - 68)^{0.33}(1.45)(T_s - 68) + (0.29)\left(\frac{T_s - 68}{0.82}\right)^{0.25}(4.04)(T_s - 68)$$

Simplifying,

$$682.6 = (0.319)(T_s - 68)^{1.33} + (1.23)(T_s - 68)^{1.25}$$

By trial and error,

$$T_s \approx 190°\text{F}$$

Converting to °C,

$$T_s = \left(\frac{5}{9}\right)(190 - 32) = \boxed{87.8°\text{C}} \quad \left[\text{answer for part (b)}\right]$$

$$h_{\text{top}} = (0.22)(190 - 68)^{0.33} = 1.074$$

$$h_{\text{sides}} = (0.29)\left(\frac{190 - 68}{0.82}\right)^{0.25} = 1.01$$

$$q_{\text{top}} = (1.074)(1.45)(190 - 68) = 190 \text{ BTU/hr}$$

$$q_{\text{sides}} = (1.01)(4.04)(190 - 68) = 498 \text{ BTU/hr}$$

$$q_{\text{total}} = 190 + 498 = 688 \text{ BTU/hr} \quad \left[\text{close enough to 682}\right]$$

The heat loss from the sides and top is

$$q_{\text{sides}} = \left(\frac{498}{688}\right)(200 \text{ W}) = \boxed{144 \text{ W}} \quad \left[\text{answer for part (a)}\right]$$

$$q_{\text{top}} = \left(\frac{190}{688}\right)(200 \text{ W}) = \boxed{55.2 \text{ W}} \quad \left[\text{answer for part (b)}\right]$$

(c) Converting the film coefficient to English units,

$$h = \left(4.883 \ \frac{\text{kcal}}{\text{hr-m}^2\text{-°C}}\right)\left(0.2048 \ \frac{\text{BTU-m}^2\text{-°C}}{\text{ft}^2\text{-°F-kcal}}\right) = 1.0 \text{ BTU/hr-ft}^2\text{-°F}$$

$$A_{\text{total}} = 1.45 + 4.04 = 5.49 \text{ ft}^2$$

$$q = hA(T_s - T_\infty + \sigma F_e F_a A)(T_s^4 - T_\infty^4)$$

$$682.6 = (1)(5.49)(T_s - 68) + (0.1713 \times 10^{-8})(0.7)(1)(5.49) \times \left[(T_s + 460)^4 - (68 + 460)^4\right]$$

Simplifying,

$$682.6 = (5.49)(T_s - 68) + (6.58 \times 10^{-9}) \times \left[(T_s + 460)^4 - (528)^4\right]$$

By trial and error,

$$T_s \approx 135°\text{F} = \boxed{57.2°\text{C}}$$

9. Since two T_{out} values are unknown, this is an NTU (heat exchanger effectiveness) problem.

Assume $c_p = 1.0$ BTU/lbm-°F for water.

$$C_{\text{water}} = \dot{m}c_p = (12{,}000)(1) = 12{,}000 \text{ BTU/hr}$$

$$C_{\text{oil}} = (20{,}000)(0.60) = 12{,}000 \text{ BTU/hr}$$

$$\frac{C_{\text{minimum}}}{C_{\text{maximum}}} = \frac{12{,}000}{12{,}000} = 1.0$$

$$\text{NTU} = \frac{(280)(40)}{12{,}000} = 0.93$$

From an NTU-effectiveness chart,

$$\mathcal{E} \approx 0.45$$

$$0.45 = \frac{(12{,}000)(280 - T_{\text{oil,out}})}{(12{,}000)(280 - 60)}$$

$$T_{\text{oil,out}} = \boxed{181°\text{F}}$$

Since $C_{\text{minimum}}/C_{\text{maximum}} = 1.0$,

$$0.45 = \frac{(12{,}000)(T_{\text{water,out}} - 60)}{(12{,}000)(280 - 60)}$$

$$T_{\text{water,out}} = \boxed{159°\text{F}}$$

To check:

$$q_{\text{water}} = mc_p\Delta T = (12{,}000)(1)(159 - 60) = 1.188 \times 10^6 \text{ BTU/hr}$$

$$q_{\text{oil}} = (20{,}000)(0.60)(280 - 181) = 1.188 \times 10^6 \text{ BTU/hr} \quad [\text{ok}]$$

10. Assumptions:

- no radiation losses
- one-dimensional heat loss
- schedule-40 pipe
- h_{top} and h_{sides} are different

Gather data:

$$k_{\text{steel}} \approx 26.5 \text{ BTU-ft/hr-ft}^2\text{-°F}$$

$$k_{\text{concrete}} \approx 0.54 \text{ BTU-ft/hr-ft}^2\text{-°F}$$

(a) Horizontal top surface:

$$A = (12)(12) = 144 \text{ ft}^2$$
$$L = 12 \text{ ft}$$

Assume $T_{s,\text{steel}} = 10°\text{F}$.

$$\Delta T = T_s - T_\infty = 10 - (-10) = 20°\text{F}$$
$$T_{\text{film}} = \left(\tfrac{1}{2}\right)(T_s + T_\infty) = \left(\tfrac{1}{2}\right)[10 + (-10)] = 0°\text{F}$$

At 0°F,

$$\text{Pr} = 0.73$$
$$\frac{g\beta\rho^2}{\mu^2} = 4.2 \times 10^6$$
$$\text{Gr} = (12)^3(4.2 \times 10^6)(42) = 3.05 \times 10^{11}$$
$$\text{PrGr} = (0.73)(3.05 \times 10^{11}) = 2.2 \times 10^{11}$$

Use a correlation for horizontal plates even though PrGr is out of range since this is an area where correlations are lacking.

$$h_{\text{top}} = (0.22)(20)^{0.33} = 0.591$$

The total resistance to heat flow is

$$R_{\text{total,top}} = \frac{1}{0.591} + \frac{\frac{1}{12}}{26.5} + \frac{\frac{14}{12}}{0.54} + \frac{1}{500} = 3.86$$
$$q_{\text{top}} = \frac{A\Delta T}{R_{\text{total}}} = \frac{(144)[50 - (-10)]}{3.86} = 2238 \text{ BTU/hr}$$

Vertical sides:

$$A = (4)(144) = 576 \text{ ft}^2$$
$$L = 12 \text{ ft}$$
$$\Delta T = 20°\text{F}$$
$$\text{PrGr} = 2.2 \times 10^{11} \quad \text{[no change]}$$
$$h = (0.19)(20)^{0.33} = 0.51$$
$$R_{\text{total,sides}} = \frac{1}{0.51} + \frac{\frac{1}{12}}{26.5} + \frac{\frac{14}{12}}{0.54} + \frac{1}{500} = 4.13$$
$$q_{\text{sides}} = \frac{(576)[50 - (-10)]}{4.13} = 8368 \text{ BTU/hr}$$

The heat required to offset the loss is

$$q_{\text{total}} = q_{\text{top}} + q_{\text{sides}} = 2238 + 8368 = \boxed{10{,}606 \text{ BTU/hr}}$$

Check the 10°F assumption about surface temperature.

$$R_{\text{to inside film}} = \frac{\frac{1}{12}}{26.5} + \frac{\frac{14}{12}}{0.54} + \frac{1}{500} = 2.17$$
$$q = \frac{A\Delta T}{R}$$
$$2238 = \frac{(144)(50 - T_s)}{2.17}$$
$$T_s = 16.3°\text{F} \quad \text{[close enough]}$$

(Checking T_s for the sides would be similar.)

(b)

- T_s would be smaller.
- T_{film} would be smaller.
- ΔT in the convection equation would be smaller.
- The heat required might include all or part of the latent heat of fusion, depending on how (in what phase) the liquid was maintained.

(c) The wind would increase h. Typically, h is calculated for v = 0 and then corrected for velocity.

(d)

$$5 \text{ psig} \approx 20 \text{ psia}$$
$$T_{\text{steam}} = 227.96°\text{F}$$
$$D_o = \frac{1.315}{12} = 0.1096 \text{ ft}$$

The area per unit length of pipe is

$$A = \pi(0.1096)(1) = 0.3443 \text{ ft}^2/\text{ft}$$

Assume $h_i = 2000$ (condensing vapor) and $h_o = 500$. Ignore steel pipe resistance.

$$\frac{1}{U_o} = \frac{1}{h_i} + \frac{1}{h_o} = \frac{1}{2000} + \frac{1}{500}$$

$U_o \approx 400$ (accurate enough since h_i and h_o were approximated.) Then,

$$q = UA\Delta T$$
$$10{,}606 = (400)(0.3443)L(227.96 - 50)$$
$$L = \boxed{0.43 \text{ ft} \quad \text{[seems low]}}$$

11. $q = (25)(3.413) = 85.3$ BTU/hr

$$T_\infty = 60°\text{F} \quad [520°\text{R}]$$

$$A = \frac{(5 \text{ sides})(5)(5)}{144} = 0.868 \text{ ft}^2$$

$$q = hA(T_s - T_\infty) + \sigma F_e F_a A(T_s^4 - T_\infty^4)$$

For T_s in °R,

$$\begin{aligned} 85.3 = &(1.5)(0.868)(T_s - 520) \\ &+ (0.1713 \times 10^{-8})(0.92)(1)(0.868) \\ &\times \left[T_s^4 - (520)^4\right] \end{aligned}$$

Simplifying,

$$65.5 = (T_s - 520) + (1.05 \times 10^{-9}) \left[T_s^4 - (520)^4\right]$$

By trial and error,

$$T_s = 560°\text{R} \quad [100°\text{F}]$$

The source box will not reach 150°F.

HEATING, VENTILATING, AND AND AIR CONDITIONING

1. This is an adiabatic saturation process.

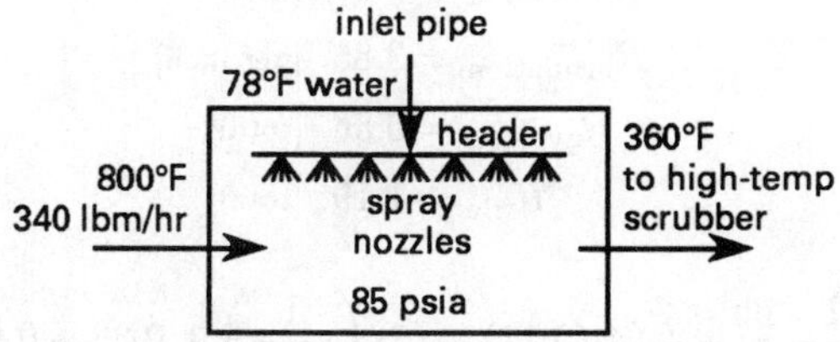

(a) For the air:

$$800°\text{F} = 1260°\text{R}$$
$$h_{\text{air,in}} = 306.65 \text{ BTU/lbm}$$
$$360°\text{F} = 820°\text{R}$$
$$h_{\text{air,out}} = 196.69 \text{ BTU/lbm}$$

For the water:

At 78°F,

$$h_{\text{water,in}} = 46.02 \text{ BTU/lbm}$$

For the water vapor:

h_{vapor} should be evaluated at p_{vapor}, which is unknown. Actually, h does not depend much on p, so assume that $p_{\text{vapor}} \approx 10$ psia.

At 360°F and 10 psia,

$$h_{\text{vapor,out}} \approx 1222 \text{ BTU/lbm}$$

Assume an adiabatic process, so the air supplies the heat of vaporization.

$$q_{\text{out,air}} = q_{\text{in,water}}$$
$$(340)(306.6 - 196.69) = \dot{m}_{\text{water}}(1222 - 46.02)$$
$$\dot{m}_{\text{water}} = \boxed{31.78 \text{ lbm/hr}}$$

(b)
$$\phi = \frac{p_{\text{water}}}{p_{\text{saturated}}}$$

From the steam tables, $p_{\text{saturated,360 °F}} = 153.04$ psia. The partial pressure is proportional to the mole fraction, x.

$$x_i = \frac{n_i}{\sum n_i}$$

The number of moles of water per hour is

$$n_{\text{water}} = \frac{31.78}{18} = 1.77$$
$$\text{MW}_{\text{air}} = 29$$
$$n_{\text{air}} = \frac{340}{29} = 11.72 \text{ moles}$$

The mole fraction is

$$x_{\text{water}} = \frac{1.77}{1.77 + 11.72} = 0.131$$

The water partial pressure is

$$p_{\text{water}} = (0.131)(85) = 11.14 \text{ psia}$$

The relative humidity is

$$\phi = \frac{11.14}{153.04} = \boxed{0.073 \quad [7.3\%]}$$

2. Find Δq from conduction.

$$\Delta T = 70 - 58 = 12°\text{F}$$
$$q = UA\Delta T = (0.15)(10{,}000)(12) + (1.10)(2500)(12) + (0.06)(25{,}000)(12) + (1.6)(720)(12)$$
$$= 82{,}824 \text{ BTU/hr}$$

Find q from infiltration. (The humidity data is not known, so Δh cannot be determined.)

$$q = (0.018)(Q)\Delta T$$
$$= (0.018)\left[(0.5)(600{,}000)\right](12)$$
$$= 64{,}800 \text{ BTU/hr}$$

The total heat loss is

$$q_{\text{total}} = 82{,}824 + 64{,}800 = 147{,}624 \text{ BTU/hr}$$

The number of unoccupied hours per heating season is

$$\left(23\ \frac{\text{wk}}{\text{yr}}\right)\left[\left(14\ \frac{\text{hr}}{\text{day}}\right)\left(5\ \frac{\text{day}}{\text{week}}\right) + \left(24\ \frac{\text{hr}}{\text{day}}\right)\left(2\ \frac{\text{day}}{\text{week}}\right)\right] = 2714 \text{ hr/yr}$$

The energy savings is

$$\frac{\left(2714\ \frac{\text{hr}}{\text{yr}}\right)\left(147{,}624\ \frac{\text{BTU}}{\text{hr}}\right)}{\left(1 \times 10^5\ \frac{\text{BTU}}{\text{therm}}\right)(0.77)} = 5203 \text{ therms}$$

The savings is

$$(0.30)(5203) = \boxed{\$1561}$$

This assumes instantaneous heating upon start-up and ventilation during unoccupied time.

3. This is a heat gain (summertime) problem.

Thermal delay will occur but is beyond the scope of a hand solution. Use the total equivalent temperature difference concept.

$$q = UA\Delta T_{te}$$

The daily swing is 24°, so the correction to ΔT_{te} is

$$(24 - 20)\left(\tfrac{1}{2}\right) = +2°$$

The difference in design temperatures is 96 − 78, so the correction to ΔT_{te} is

$$(96 - 78) - 20 = -2°$$

The total correction is +2 − 2 = 0°F.

(a) Develop the U values.

$$U = \frac{1}{\sum \frac{1}{h} + \sum \frac{x}{k}} = \frac{1}{\sum \frac{1}{h} + \sum R_{th}}$$

Walls:

$$\begin{aligned} h_i &= 1.65 \quad \text{[assumed]} \\ h_o &= 4.0 \quad \text{[assumed, 7.5 mph]} \\ R_{th,\text{4-in brick facing}} &= 0.44 \quad \text{[total]} \\ R_{th,\text{4-in clay tile}} &= 1.11 \quad \text{[total]} \\ R_{th,\text{insulation}} &= 3.85 \quad \text{[per inch]} \\ R_{\text{5/8-in gypsum}} &= 0.56 \quad \text{[total]} \end{aligned}$$

$$U_{walls} = \frac{1}{\frac{1}{4} + 0.44 + 1.11 + (3.5)(3.85) + 0.56 + \frac{1}{1.65}} = 0.061 \text{ BTU/hr-ft}^2\text{-°F}$$

Obtain the values of ΔT_{te}.

facing	ΔT_{te}
N	17
S	28
E	20
W	20

The heat gain is

$$\begin{aligned} q &= UA\Delta T_{te} \\ &= (0.061)\left[(1700)(17) + (1500)(28) + (1600)(20) + (1500)(20)\right] \\ &= 8107 \text{ BTU/hr} \end{aligned}$$

Roof:

Assume 4-ply means $\frac{3}{8}$-in built-up roofing.

$$\begin{aligned} R_{concrete} &= 0.11 \quad \text{[per inch]} \\ R_{\text{built-up roof}} &= 0.33 \quad \text{[total]} \\ R_{plywood} &= 0.63 \quad \text{[total]} \\ R_{insulation} &= 3.85 \quad \text{[per inch]} \\ R_{gypsum} &= 0.56 \quad \text{[total]} \\ R_{tile} &= 1.19 \quad \text{[total]} \end{aligned}$$

$$\begin{aligned} \frac{1}{U_{roof}} &= \tfrac{1}{4} + (4)(0.11) + (2)\left(\frac{1}{0.28}\right) + 0.33 + 0.63 \\ &\quad + (8.5)(3.85) + 0.56 + 1.19 + \frac{1}{1.65} \\ &= 43.87 \\ U &= \frac{1}{43.87} = 0.023 \\ \Delta T_{te} &= 74\text{°F} \\ q_{roof} &= (0.023)(6000)(74) = 10{,}212 \text{ BTU/hr} \end{aligned}$$

Windows:

$$\begin{aligned} k &= 0.45 \\ U &= \frac{1}{\frac{1}{4} + \frac{\frac{0.25}{12}}{0.45} + \frac{1}{1.65}} \\ &= 1.11 \end{aligned}$$

The light-colored venetian blinds require a factor of 0.56 (from ASHRAE or Carrier handbooks) to be applied to solar heat gain.

$$\begin{aligned} q &= A\left[C_s F_{shg} + U(T_o - T_i)\right] \\ C_s &= 0.95 \text{ for } \tfrac{1}{4}\text{-in single plate glass} \\ F_{shg} &= 26 \quad \text{[use 32° north latitude table]} \end{aligned}$$

Note that $F_{shg} \neq 0$ even though the sun is in the west and the windows face east. The residual gain is from diffuse radiation.

$$\begin{aligned} q &= (150)\left[(0.95)(0.56)(26) + (1.11)(96 - 78)\right] \\ &= 5072 \text{ BTU/hr} \end{aligned}$$

The total instantaneous heat gain is

$$\begin{aligned} q_{total} &= q_{walls} + q_{roof} + q_{windows} \\ &= 8107 + 10{,}212 + 5072 \\ &= \boxed{23{,}391 \text{ BTU/hr}} \end{aligned}$$

(b) This may or may not be the peak cooling load. Both F_{shg} and ΔT_{te} peak at other times of day. A trial and error determination is needed.

4. Solve this as a recirculating air bypass problem. (It can also be solved without bypass, and the solution is slightly simpler.)

Assume the metabolic heat per person is 330 BTU/hr total (225 BTU/hr sensible and 105 BTU/hr latent). The answer will vary depending on the values chosen.

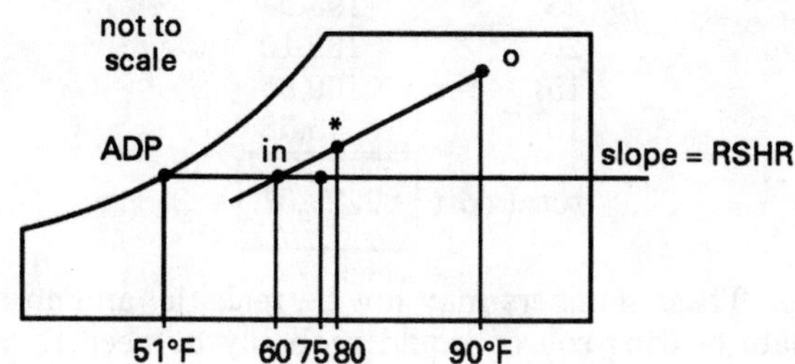

step 1:

$$h_i = 28.1 \text{ BTU/lbm}$$
$$h_o = 33.9 \text{ BTU/lbm}$$
$$v_o = 14.1 \text{ ft}^3\text{/lbm}$$

step 2:

$$q_{\text{sensible}} = (600)(225) + 100{,}000 = 235{,}000 \text{ BTU/hr}$$
$$q_{\text{latent}} = (600)(105) = 63{,}000 \text{ BTU/hr}$$
$$\text{RSHR} = \frac{235{,}000}{235{,}000 + 63{,}000} = 0.79$$

step 3: The apparatus dew point is

$$\text{ADP} \approx \boxed{51°\text{F}}$$

step 5: Select $T_{\text{in}} = 60°\text{F}$ (given). At point "in":

$$T_{\text{db}} = 60°\text{F}$$
$$T_{\text{wb}} = 55.6°\text{F}$$
$$\omega = 52.5 \text{ grains}$$

step 6: Q_{in} is limited by sensible heat removal.

$$Q_{\text{in}} = \frac{(55.3)(235{,}000)}{75 - 60} = 8.66 \times 10^5 \text{ ft}^3\text{/hr}$$

(Q_{in} is not limited by latent heat removal since $\Delta\omega$ is small.)

$$Q_{\text{in}} = \frac{8.66 \times 10^5}{60} = \boxed{14{,}433\ (14{,}000)\ \text{ft}^3\text{/min}}$$

step 7: Determine the system bypass factor. Work with dry bulb temperatures.

$$\text{BF}_{\text{system}} = \frac{60 - 51}{75 - 51} = 0.375$$
$$Q_2 = (\text{BF})(Q_{\text{in}}) = (0.375)(8.66 \times 10^5) = 324{,}750 \text{ ft}^3\text{/hr}$$
$$Q_1 = 8.66 \times 10^5 - 324{,}750 = 541{,}250 \text{ ft}^3\text{/hr}$$

step 8: Outside air is

$$Q_o = (600)(5)(60) = 180{,}000 \text{ ft}^3\text{/hr}$$

Locating point * based on dry bulb temperatures,

$$\frac{Q_o}{Q_1} = \frac{180{,}000}{541{,}250} = 0.333 \quad \text{[1/3 outside air]}$$
$$T_{*,\text{db}} = T_{i,\text{db}} + (0.333)(T_{o,\text{db}} - T_{i,\text{db}}) = 75 + (0.333)(90 - 75) = \boxed{80°\text{F entering coil}}$$

step 11:

$$q_t = 235{,}000 + 63{,}000 + (33.9 - 28.1)(180{,}000)\left(\frac{1}{14.1}\right) \approx 372{,}000 \text{ BTU/hr}$$

The tonnage is

$$\frac{372{,}000}{12{,}000} = \boxed{31 \text{ tons}}$$

From $q = \dot{m}c_p\Delta T$,

$$\dot{m}_{\text{water}} = \frac{372{,}000}{(1)(52° - 42°)} = \boxed{37{,}200 \text{ lbm/hr}}$$

5. (a)

$$q_{\text{design}} = UA(T_{\text{id}} - T_{\text{od}})$$
$$3 \times 10^5 = UA(70 - 0)$$
$$UA \approx 4290 \text{ BTU/hr-°F}$$

At 65°F,

$$q_{65°\text{F}} = (4290)(70 - 65) = 21{,}450$$

The heat pump is adequate at 65°F.

At 60°F,

$$q_{60^\circ\text{F}} = (4290)(70 - 60) = 42{,}900$$

The heat pump is adequate.

At 55°F,

$$\begin{aligned} q_{55\ ^\circ\text{F}} &= (4290)(70 - 55) = 64{,}350 \\ q_{\text{electrical}} &= 64{,}350 - 50{,}000 - 7500 \\ &= 6850\ \text{BTU/hr} \\ \text{cost} &= \left(6850\ \frac{\text{BTU}}{\text{hr}}\right)(450\ \text{hr}) \\ &\quad \times \left(2.93 \times 10^{-4}\ \frac{\text{kW-hr}}{\text{BTU}}\right)\left(0.08\ \frac{\$}{\text{kW-hr}}\right) \\ &= \$72.25\ \ [\$72] \end{aligned}$$

The following table is prepared in the same manner.

T	q_t	$q_{\text{electrical}}$	cost
65	21,450	0	0
60	42,900	0	0
55	64,350	6850	72
50	85,800	33,300	265
45	107,250	59,750	350
40	128,700	86,200	384
35	150,150	112,650	343
30	171,600	139,100	293
25	193,050	165,600	233
20	214,500	192,000	135
15	235,950	218,450	77
10	257,400	244,900	17
		annual total cost	**$2169**

(b) At 65°F, the capacity of the heat pump exceeds the loss, so the heat pump will run only part time. The cost of this part-time operation is

$$\left(\$0.08\ \frac{1}{\text{kW-hr}}\right)(6\ \text{kW})(820\ \text{hr}) \left(\frac{21{,}450\ \frac{\text{BTU}}{\text{hr}} - 7500\ \frac{\text{BTU}}{\text{hr}}}{60{,}000\ \frac{\text{BTU}}{\text{hr}}}\right) = \$91.51$$

At 60°F, the heat pump will also run only part time, but the capacity of the heat pump is lower. The cost is

$$(\$0.08)(5.8\ \text{kW})(600\ \text{hr})\left(\frac{42{,}900 - 7500}{55{,}000}\right) = \$179.19$$

The following amounts are calculated similarly.

T	cost
65	91.51
60	179.19
55	229.22
50	255.57
45	259.35
40	263.18
35	237.37
30	217.40
25	195.94
20	139.10
15	109.66
10	45.58
total cost	**$2223.07**

Note: These numbers may not be realistic, and an alternate heating source would probably be needed. At the higher temperatures, the heat pump only has to run a fraction of an hour to replace the heat lost each hour. However, at the lower temperatures, the heat pump's lower capacity means that it is running continuously. For example, the heat pump must run almost 50 hr for each hour spent at 10°F.

6. This is a standard setback problem.

From $q = UA\Delta T$, the heat loss per °F is

$$\frac{q}{\Delta T} = UA = \frac{135{,}000}{70} = 1928\ \text{BTU/hr-°F}$$

The overall average outdoor temperature, $\overline{\overline{T}}$, during the heating season is

$$\begin{aligned} \text{DD} &= N(65 - \overline{\overline{T}}) \\ 4839 &= (220)(65 - \overline{\overline{T}}) \\ \overline{\overline{T}} &= 43^\circ\text{F} \end{aligned}$$

The original heat loss during the heating season was

$$\begin{aligned} q &= UA\Delta T \\ &= (1928)(73 - 43)\left(24\ \frac{\text{hr}}{\text{day}}\right) \\ &\quad \times \left(220\ \frac{\text{day}}{\text{heating season}}\right) \\ &= 3.05 \times 10^8\ \text{BTU} \end{aligned}$$

The setback heat loss is

$$\begin{aligned} q &= (1928)(68 - 43)(24)(220) \\ &= 2.54 \times 10^8\ \text{BTU} \end{aligned}$$

The savings is

$$3.05 \times 10^8 - 2.54 \times 10^8 = \boxed{5.1 \times 10^7\ \text{BTU/yr}}$$

7. This is a recirculating air bypass problem.

Assume the following metabolic heating in BTU/hr (other values are possible).

	seated	dancing
male	390	1000
female*	330	850
50%–50%, average	360	925

* 85% of male value

	seated	dancing	average
sensible	250	345	297.5 [say 300]
latent	110	580	345 [say 350]
total	360	925	

step 1:

$$h_i = 29.3 \text{ BTU/lbm}$$
$$h_o = 39.4 \text{ BTU/lbm}$$
$$v_o = 14.2 \text{ ft}^3\text{/lbm}$$

step 2:

$$q_s = (240)(300) + 140{,}000 = 212{,}000 \text{ BTU/hr}$$

Assume $h_{FG} \approx 1060$ BTU/lbm (corresponds to 60°F).

$$q_e = (240)(350) + \left(\frac{45{,}000}{7000}\right)(1060) = 90{,}800$$

$$\text{RSHR} = \frac{212{,}000}{212{,}000 + 90{,}800} = 0.70$$

step 3: From the psychrometric chart, ADP ≈ 48°F.

step 5:

$$T_{\text{in}} = 60°\text{F} \quad \text{[given]}$$
$$\omega \approx \boxed{60 \text{ grains/lbm}}$$

step 6:

$$Q_{\text{in}} = \frac{(55.3)(212{,}000)}{77 - 60} = \boxed{6.90 \times 10^5 \text{ ft}^3\text{/hr}}$$

step 8:

$$Q_o = (4500)(60) = 270{,}000 \text{ ft}^3\text{/hr}$$

step 11:

$$q_t = \frac{212{,}000 + 90{,}800 + (39.4 - 29.3)(270{,}000)\left(\frac{1}{14.2}\right)}{12{,}000 \ \frac{\text{BTU}}{\text{hr-ton}}} = \boxed{41.2 \text{ tons}}$$

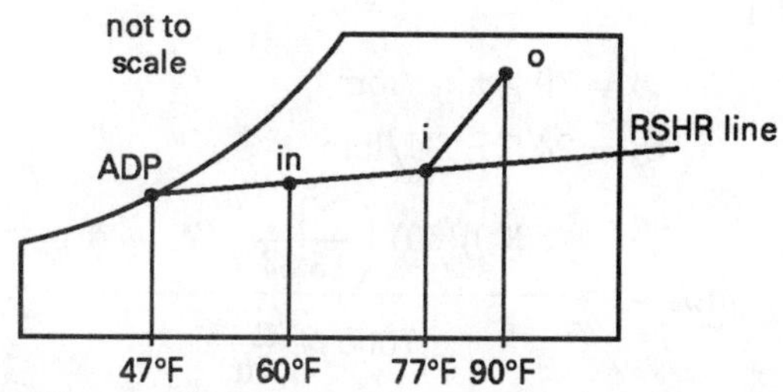

8. (a) The room is maintained at 75°F and 40%. From the psychrometric chart,

$$v_o = 13.64 \text{ ft}^3\text{/lbm}$$
$$\omega_i = 52 \text{ grains/lbm} \quad [0.0074 \text{ lbm/lbm}]$$

From a low-temperature psychrometric chart,

$$v_o = 12.25 \text{ ft}^3\text{/lbm}$$
$$\omega_o = 0.00055 \text{ lbm/lbm}$$

The moisture loss is

$$(800)(60)\left[\left(\frac{1}{13.64}\right)(0.0074) - \left(\frac{1}{12.25}\right)(0.00055)\right] = 23.9 \text{ lbm/hr}$$

Assume the latent load is approximately 200 BTU/hr-person.

Assume $h_{FG} \approx 1052$ BTU/lbm (corresponds to 75°F).

The metabolic moisture gain is

$$\frac{(35)(200)}{1052} = 6.7 \text{ lbm/hr}$$

The humidifier must supply

$$23.9 - 6.7 = \boxed{17.2 \text{ lbm/hr}}$$

(b) The dew point at the interior conditions is

$$T_{\text{dp}} \approx 49°\text{F}$$

If the interior of the windows is below 49°F, which is likely, condensation will occur.

$$\boxed{\text{Yes}}$$

9. At 80°F,

$$\omega_1 = 77 \text{ grains/lbm}$$
$$v_1 = 13.84 \text{ ft}^3\text{/lbm}$$
$$h_1 = 31.2 \text{ BTU/lbm}$$

At 57°F,

$$\omega_2 = 70 \text{ grains/lbm}$$
$$h_2 = 24.5 \text{ BTU/lbm}$$
$$\dot{m}_{\text{water}} = \frac{(8000)(60)\left(\frac{1}{13.84}\right)(77-70)}{7000 \frac{\text{grains}}{\text{lbm}}}$$
$$= \boxed{34.7 \text{ lbm/hr} \quad \text{[removed]}}$$

The refrigeration effect is calculated for air.

$$q_t = (8000)(60)\left(\frac{1}{13.84}\right)(31.2 - 24.5)$$
$$= 2.32 \times 10^5 \text{ BTU/hr}$$
$$= \frac{2.32 \times 10^5}{12{,}000} = \boxed{19.3 \text{ tons}}$$

The chilled water flow rate is

$$\dot{m} = \frac{q_t}{c_p \Delta T} = \frac{2.32 \times 10^5}{(1)(55-42)}$$
$$= 17{,}846 \text{ lbm/hr}$$

Converting to gpm,

$$\frac{\left(17{,}846 \frac{\text{lbm}}{\text{hr}}\right)\left(7.48 \frac{\text{gal}}{\text{ft}^3}\right)}{\left(62.4 \frac{\text{lbm}}{\text{ft}^3}\right)\left(60 \frac{\text{min}}{\text{hr}}\right)} = \boxed{35.7 \text{ gpm}}$$

10.

$$q_s = 70{,}000 + 15{,}000 + (3000)(3)(4.5)$$
$$= 125{,}500 \text{ BTU/hr}$$
$$q_l = (3000)(3)(2) = 18{,}000 \text{ BTU/hr}$$
$$\text{RSHR} = \frac{125{,}500}{125{,}500 + 18{,}000} = 0.87$$

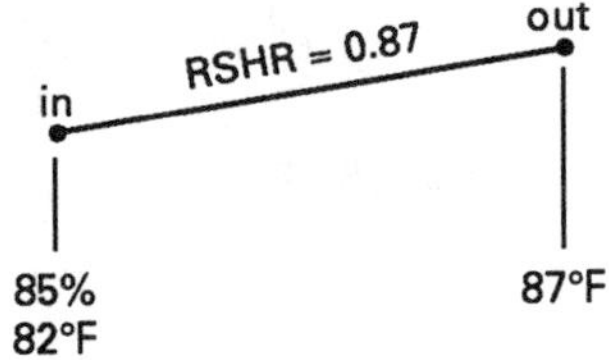

At the out point,

$$T_{\text{out}} = 87°\text{F}$$
$$\omega \approx \boxed{144 \text{ grains/lbm} \ [0.02057 \text{ lbm/lbm}]}$$
$$Q = \frac{125{,}500}{(0.018)(60)(87-82)} = \boxed{23{,}240 \text{ ft}^3/\text{min}}$$
$$A = \frac{Q}{V} = \frac{23{,}240}{140} = \boxed{166 \text{ ft}^2}$$

11.

$$\rho_{\text{air}} = \frac{p}{RT} = \frac{(14.7)(144)}{(53.3)(460+60)}$$
$$= 0.07637 \text{ lbm/ft}^3$$
$$m_{\text{air}} = Q\rho = (20{,}000)(0.07637)$$
$$= 1527 \text{ lbm/hr}$$

Suction conditions:

$$p_t = 13.4 \text{ psia} = p_{\text{air}} + p_{\text{water}}$$

Since $\phi = 0.80 = p_{\text{water}}/p_{\text{saturated}}$, and $p_{\text{saturated}} = 0.5959$ psia at 85°F,

$$p_{\text{water}} = (0.80)(0.5959) = 0.477 \text{ psia}$$
$$p_{\text{air}} = 13.4 - 0.477 = 12.923 \text{ psia}$$

Take 100 ft^3 of air at suction conditions.

$$m_{\text{air}} = \frac{p_{\text{air}} V}{R_{\text{air}} T} = \frac{(12.923)(144)(100)}{(53.3)(460+85)}$$
$$= 6.406 \text{ lbm}$$
$$m_{\text{water}} = \frac{p_{\text{water}} V}{R_{\text{water}} T} = \frac{(0.477)(144)(100)}{(85.8)(460+85)}$$
$$= 0.147 \text{ lbm}$$

The suction specific humidity is

$$\omega = \frac{m_{\text{water}}}{m_{\text{air}}} = \frac{0.147}{6.406} = \boxed{0.0229 \text{ lbm/lbm}}$$

Discharge conditions:

Since $p_2 = p_1(T_2/T_1)$ for a constant-volume process,

$$p = (100)\left(\frac{75+460}{500+460}\right) = 55.73 \text{ psia}$$
$$p_{\text{water}} = 0.4298 \quad \text{[saturated at 75°F]}$$
$$p_{\text{air}} = p_t - p_{\text{water}} = 55.73 - 0.4298 \approx 55.3 \text{ psia}$$
$$\omega = \frac{m_{\text{water}}}{m_{\text{air}}} = \frac{\frac{p_{\text{water}} V}{R_{\text{water}} T}}{\frac{p_{\text{air}} V}{R_{\text{air}} T}} = \frac{p_{\text{water}} R_{\text{air}}}{p_{\text{air}} R_{\text{water}}}$$
$$= \frac{(0.4298)(53.3)}{(55.3)(85.8)}$$
$$= \boxed{0.00483 \text{ lbm/lbm}}$$

The water removed is

$$m_{\text{water}} = m_{\text{air}} \Delta\omega$$
$$= (1527)(0.0229 - 0.00483)$$
$$= \boxed{27.59 \text{ lbm/hr}}$$

12. (a) Assume hot plate load is $\frac{2}{3}$ sensible, $\frac{1}{3}$ latent.

Sensible heat gain:

Occupants:	$(2)(225) = 450$ BTU/hr	
Lights:	$(3)(500)(3.413) = 5120$ BTU/hr	
Hot plates:	$\left(\frac{2}{3}\right)(18)(500)(3.413) = 20{,}480$ BTU/hr	
	Total sensible load:	**26,050 BTU/hr**

Latent heat gain:

Occupants:	$(2)(105) = 210$ BTU/hr	
Hot plates:	$\left(\frac{1}{3}\right)(18)(500)(3.413) = 10{,}240$ BTU/hr	
	Total latent load:	**10,450 BTU/hr**

(b) $q = \dot{m}c_p\Delta T$ and $\dot{m} = Q\rho$.

$$\dot{m}_{\text{air}} = \frac{q}{c_p\Delta T} = \frac{26{,}050}{(0.241)(16)} = 6756 \text{ lbm/hr}$$

$$Q_{\text{in}} = \frac{\dot{m}}{\rho} \approx \frac{6756}{0.075} = \boxed{90{,}080 \text{ ft}^3/\text{hr}}$$

The need for greater accuracy in ρ is questionable.

(c) $\text{SHR} = \dfrac{26{,}050}{26{,}050 + 10{,}450} = 0.71$

The "in" condition point is found as follows.

(1) Locate the "out" point on the psychrometric chart.

(2) Draw a line with a slope of SHR = 0.71 through the "out" point.

(3) The intersection of the SHR line and $T_{\text{db}} = 72 - 16 = 56$ is the "in" point.

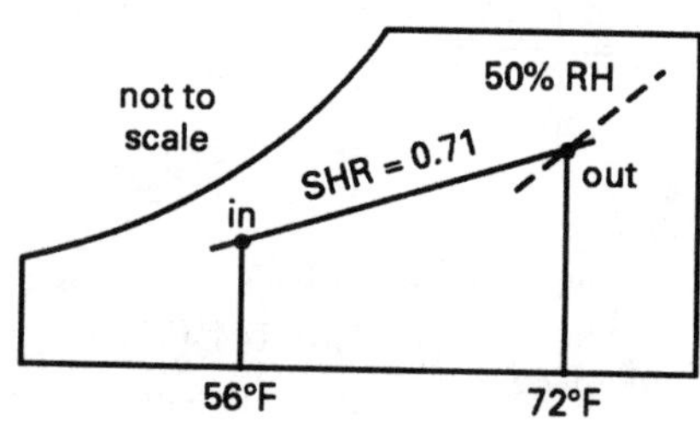

$$T_{\text{db}} = \boxed{56°\text{F}}$$

$$\phi = \boxed{72\%}$$

(d) The volume of air removed in the fume hoods is

$$(2)(1400)(60) = \boxed{168{,}000 \text{ ft}^3/\text{min}}$$

The air from the hood supply is

$$(0.65)(168{,}000) = \boxed{109{,}200 \text{ ft}^3/\text{min}}$$

The hood supply deficit (35%) is

$$168{,}000 - 109{,}200 = 58{,}800$$

$$\begin{aligned} Q_{\text{leakage}} &= (0.10)(90{,}080 + 109{,}200) \\ &= 19{,}930 \\ Q_{\text{return}} &= 109{,}200 + 90{,}080 \\ &\quad - 168{,}000 - 19{,}930 \\ &= \boxed{11{,}350} \end{aligned}$$

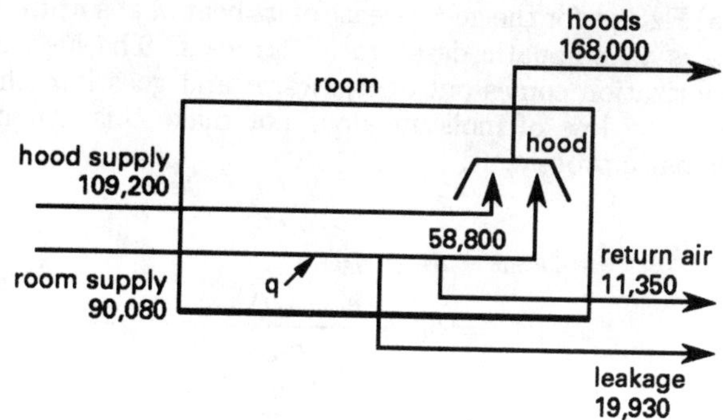

13. $\omega_{\text{out}} = \dfrac{25 \ \frac{\text{grains}}{\text{lbm}}}{7000 \ \frac{\text{grains}}{\text{lbm}}} = 0.00357 \text{ lbm/lbm}$

From the psychrometric chart,

$$\begin{aligned} \omega_{\text{in}} &= 0.008 \\ v_{\text{in}} &\approx 13.25 \text{ ft}^3/\text{lbm} \\ h_{\text{in}} &\approx 23.2 \text{ BTU/lbm} \end{aligned}$$

The air mass is

$$m_{\text{air}} = \frac{\left(15{,}000 \ \frac{\text{ft}^3}{\text{min}}\right)\left(60 \ \frac{\text{min}}{\text{hr}}\right)(5 \text{ hr})}{13.25 \ \frac{\text{ft}^3}{\text{lbm}}} = 3.4 \times 10^5 \text{ lbm}$$

(d) Since the steady-state moisture removed from the drying beds equals the moisture removed from the air,

$$\begin{aligned} m_{\text{water}} &= m_{\text{air}}\Delta\omega \\ &= (3.4 \times 10^5)(0.008 - 0.00357) \\ &= \boxed{1506 \text{ lbm}} \end{aligned}$$

(b)
$$0.05 = \frac{m_{\text{water,dry}}}{m_{\text{gel}} + m_{\text{water,dry}}}$$
$$\begin{aligned} 0.25 &= \frac{m_{\text{water,wet}}}{m_{\text{gel}} + m_{\text{water,wet}}} \\ &= \frac{m_{\text{water,dry}} + 1506}{m_{\text{gel}} + m_{\text{water,dry}} + 1506} \end{aligned}$$

Solving these two equations simultaneously,

$$m_{\text{water,dry}} = 282 \text{ lbm}$$
$$m_{\text{gel}} = \boxed{5361 \text{ lbm}}$$

Notice that the basis (i.e., the denominator) is different for these weight-basis fractions.

(a) Except for the gel's release of its heat of absorption, this is an adiabatic desaturation process. The heat of vaporization comes out of the water and goes into the air. The loss of moisture does not make this a non-adiabatic process.

$$\begin{aligned} h_{\text{out}} &= h_{\text{in}} + q_{\text{gel}} \\ &= h_{\text{in}} + \frac{m_{\text{water}}\Delta h_{\text{gel}}}{m_{\text{air}}} \\ &= 23.2 + \frac{(1506)(225)}{3.4 \times 10^5} \\ &= 24.2 \text{ BTU/lbm} \end{aligned}$$

Using h_{out} and ω_{out}, the "out" point can be located.

$$T_{\text{db,out}} = \boxed{83.5°\text{F}}$$

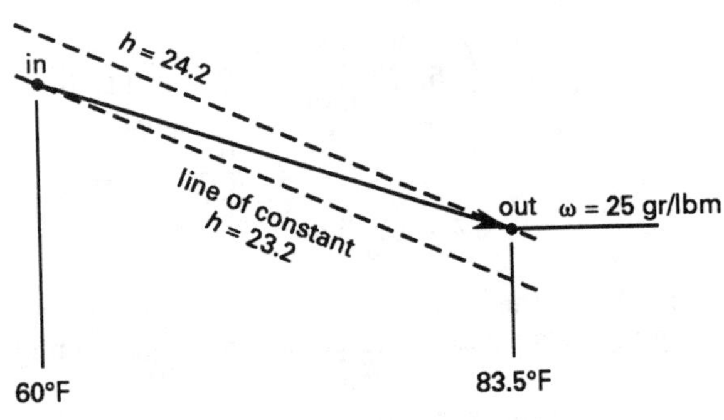

(c) The actual conditions of the air are unknown, so average values of c_p and ρ will be used.

During cooling,

$$\begin{aligned} q_{\text{gel}} &= q_{\text{air}} = mc_p\Delta T \\ (5361)(0.21)(160 - 85) &= m_{\text{air}}(0.241)(75 - 65) \\ m_{\text{air}} &= 35{,}036 \text{ lbm} \\ Q_{\text{air,cooling}} &= \frac{\dot{m}}{\rho} \\ &\approx \frac{35{,}036 \text{ lbm}}{(60 \text{ min})\left(0.075 \frac{\text{lbm}}{\text{ft}^3}\right)} \\ &= \boxed{7786 \text{ ft}^3/\text{min}} \end{aligned}$$

14.

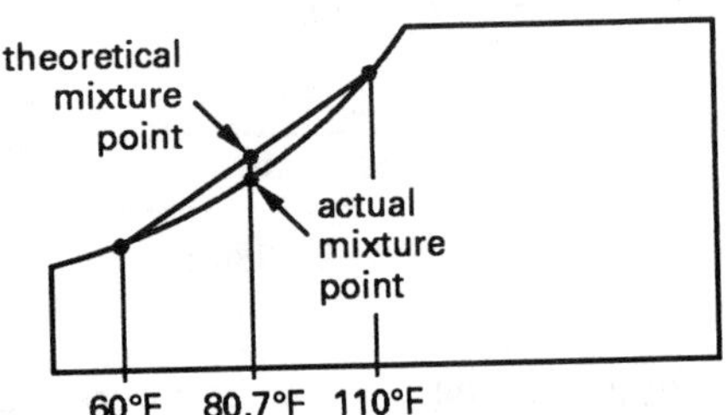

The mixture point is off the psychrometric chart, so the air will be saturated.

Stream 1:

$$\begin{aligned} p_{\text{saturated,60°F}} &= 0.2563 \text{ psia} \\ m_{\text{water,1}} &= \frac{pV}{RT} \\ &= \frac{(0.2563)(144)(1200)}{(85.8)(60 + 460)} \approx 1.0 \text{ lbm} \\ m_{\text{air,1}} &= \frac{(14.7 - 0.2563)(144)(1200)}{(53.3)(60 + 460)} \approx 90.0 \text{ lbm} \end{aligned}$$

Stream 2:

$$\begin{aligned} p_{\text{saturated,110°F}} &= 1.2748 \text{ psia} \\ m_{\text{water,2}} &= \frac{(1.2748)(144)(1000)}{(85.8)(110 + 460)} \\ &= 3.8 \text{ lbm} \\ m_{\text{air,2}} &= \frac{(14.7 - 1.2748)(144)(1000)}{(53.3)(110 + 460)} \\ &= 63.6 \text{ lbm} \end{aligned}$$

(a)
$$\begin{aligned} T &\approx \frac{(90)(60) + (63.6)(110)}{90 + 63.6} \\ &= \boxed{80.7 \quad [\text{say } 81°\text{F}]} \end{aligned}$$

If Q's are used instead of $\dot{m}$'s, $T = 82.7°F$, but this is an approximation only.

(b) For 81°F, $p_{\text{saturated}} = 0.5241$ psia (steam tables).

$$p_{\text{air}} = 14.7 - 0.5241 = 14.1759 \text{ psia}$$

$$V = \frac{mRT}{p} = \frac{(90 + 63.6)(53.3)(460 + 81)}{(14.1759)(144)} \approx 2170 \text{ ft}^3$$

At saturation,

$$m_{\text{water}} = \frac{pV}{RT} = \frac{(0.5241)(144)(2170)}{(85.8)(460 + 81)} \approx 3.5 \text{ lbm}$$

The mass of water withdrawn is

$$1.0 + 3.8 - 3.5 = \boxed{1.3 \text{ lbm [per minute]}}$$

(Alternatively, read ω_{mixture} directly and calculate $m_{\text{water}} = \Delta\omega m_{\text{air}}$.)

15. The sensible loads are

People:	(12)(225)	= 2700 BTU/hr
Lights:	(15 kW)(3413)	= 51,200 BTU/hr
Motor:		= 18,000 BTU/hr
Heat gain:		= 14,000 BTU/hr
		85,900 BTU/hr

The latent load is

People: (12)(105) = 1260 BTU/hr

(a) $Q_{\text{in}} = (4)(75{,}000) = 300{,}000 \text{ ft}^3/\text{hr}$

$$Q_{\text{in}} = \frac{(55.3)q_s}{T_i - T_{\text{in}}} = 300{,}000 = \frac{(55.3)(85{,}900)}{73 - T_{\text{in}}}$$

$$T_{\text{in}} = \boxed{57.2°\text{F}}$$

(b) The exfiltration and make-up air is

$$(0.10)(300{,}000) = 30{,}000 \text{ ft}^3/\text{hr}$$

From the psychrometric chart,

$$v_i = 13.56 \text{ ft}^3/\text{lbm}$$

$$m_{\text{air},i} = \frac{300{,}000 - 30{,}000}{13.56} = 19{,}910 \text{ lbm/hr}$$

From the psychrometric chart,

$$v_o = 14.5 \text{ ft}^3/\text{lbm}$$

$$m_{\text{air},o} = \frac{30{,}000}{14.5} = 2070 \text{ lbm/hr}$$

The mass of the air entering the coils is

$$19{,}910 + 2070 = \boxed{21{,}980 \text{ lbm/hr}}$$

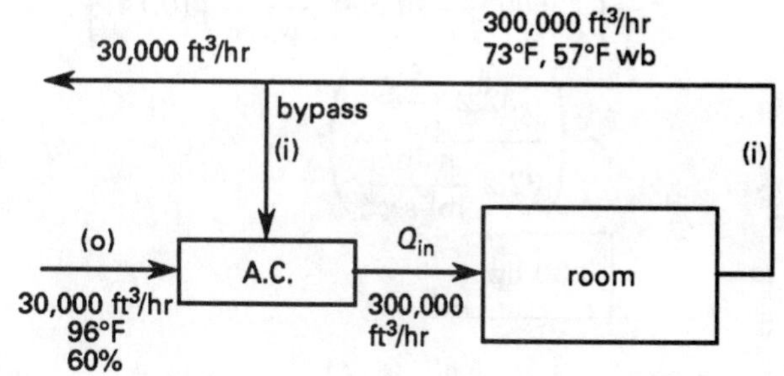

(c) $h_i = 24.3$ BTU/lbm [psychrometric chart]

$h_o = 47.6$ BTU/lbm

$q_t = q_s + q_l + (h_o - h_i)\dot{m}_{\text{air},o}$

The tonnage is

$$q_t = \frac{85{,}900 + 1260 + (47.6 - 24.3)(2070)}{12{,}000 \ \frac{\text{BTU}}{\text{hr-ton}}} = \frac{135{,}400}{12{,}000} = \boxed{11.3 \text{ tons}}$$

(d) The room sensible heat ratio is

$$\text{RSHR} = \frac{85{,}900}{85{,}900 + 1260} = 0.986 \text{ [say 1.00]}$$

This is essentially a horizontal line on the psychrometric chart.

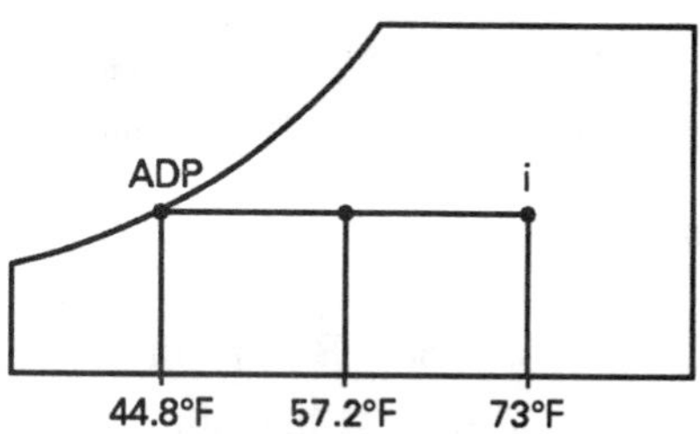

$$\text{ADP} = \boxed{44.8°\text{F}}$$

(e)
$$\dot{m} = \frac{q}{c_p\Delta T} = \frac{135{,}400}{(1)(10)} = 13{,}540 \text{ lbm/hr}$$

$$Q = \frac{\left(13{,}540 \ \frac{\text{lbm}}{\text{hr}}\right)\left(7.48 \ \frac{\text{gal}}{\text{ft}^3}\right)}{\left(62.4 \ \frac{\text{lbm}}{\text{ft}^3}\right)\left(60 \ \frac{\text{min}}{\text{hr}}\right)} = \boxed{27.1 \text{ gpm}}$$

(f) $$P = \left(\frac{\dot{m}\Delta h}{550\ \eta}\right)\left(\frac{g}{g_c}\right)$$

$$= \left[\frac{\left(13{,}540\ \frac{\text{lbm}}{\text{hr}}\right)(32\ \text{ft})}{\left(3600\ \frac{\text{sec}}{\text{hr}}\right)\left(550\ \frac{\text{ft-lbf}}{\text{hp-sec}}\right)(0.74)}\right] \times \left(\frac{32.2\ \frac{\text{ft}}{\text{sec}^2}}{32.2\ \frac{\text{ft-lbm}}{\text{lbf-sec}^2}}\right)$$

$$= \boxed{0.30\ \text{hp}}$$

16. Water:

At 120°F,

$$h = 87.92\ \text{BTU/lbm}$$
$$v = 0.01620\ \text{ft}^3/\text{lbm}$$
$$m_{\text{water}} = \frac{2750\ \text{gpm}}{\left(7.48\ \frac{\text{gal}}{\text{ft}^3}\right)\left(0.01620\ \frac{\text{ft}^3}{\text{lbm}}\right)} = 22{,}700\ \text{lbm/min}$$

At 85°F,

$$h = 53.0\ \text{BTU/lbm}$$

Air:

In:

$$h = 24.6\ \text{BTU/lbm}$$
$$v = 13.55\ \text{ft}^3/\text{lbm}$$
$$\omega = 47\ \text{grain/lbm}\quad [0.0067\ \text{lbm/lbm}]$$

Out: From a high-temperature psychrometric chart,

$$h = 63.9\ \text{BTU/lbm}$$
$$\omega = 0.037\ \text{lbm/lbm}$$

Alternatively, if a high-temperature psychrometric chart is unavailable,

$$p_{\text{saturated},100^\circ\text{F}} = 0.9492\ \text{psi}$$
$$p_{\text{water}} = (0.85)(0.9492) = 0.8068\ \text{psi}$$
$$\omega = \frac{0.622\ p_{\text{water}}}{p_{\text{total}} - p_{\text{water}}} = \frac{(0.622)(0.8068)}{14.7 - 0.8068} = 0.036\ \text{lbm/lbm}$$
$$h = (0.241)(100) + (0.036) \times [(0.444)(100) + 1061] = 63.9\ \text{BTU/lbm}$$

(a)
$$\dot{m}_{\text{water}}\Delta h_{\text{water}} = \dot{m}_{\text{air}}\Delta h_{\text{air}}$$
$$(22{,}700)(87.92 - 53.0) = m_{\text{air}}(63.9 - 24.6)$$
$$m_{\text{air}} = 20{,}170\ \text{lbm/min}$$
$$Q = (20{,}170)\left(13.55\ \frac{\text{ft}^3}{\text{lbm}}\right) = \boxed{273{,}300\ \text{ft}^3/\text{min}}$$

(b) The evaporative water loss is

$$\Delta m_{\text{water}} = m_{\text{air}}\Delta\omega = (20{,}170)(0.037 - 0.0067) = 611\ \text{lbm water/min}$$

At 53°F, $v = 0.01603\ \text{ft}^3/\text{lbm}$.

$$Q_{\text{water}} = (1.35)\left(611\ \frac{\text{lbm}}{\text{min}}\right) \times \left(0.01603\ \frac{\text{ft}^3}{\text{lbm}}\right)\left(7.48\ \frac{\text{gal}}{\text{ft}^3}\right) = \boxed{98.9\ \text{gpm}}$$

17. Outside:

$$v = 11.58\ \text{ft}^3/\text{lbm}\quad \left[\begin{array}{c}\text{low temperature}\\ \text{psychrometric chart}\end{array}\right]$$

Current (68°F):

$$h = 20.3\ \text{BTU/lbm}$$

Proposed (65°F):

$$h = 19.2\ \text{BTU/lbm}$$
$$V_{\text{room}} = (30)(60)(10) = 18{,}000\ \text{ft}^3$$
$$Q_{\text{exfiltration}} = (0.03)\left(60\ \frac{\text{min}}{\text{hr}}\right)(18{,}000\ \text{ft}^3) = 32{,}400\ \text{ft}^3/\text{hr}$$

The savings is

$$m\Delta h = Q\rho\Delta h = \frac{Q\Delta h}{v} = \frac{\left(32{,}400\ \frac{\text{ft}^3}{\text{hr}}\right)(20.3 - 19.2)}{11.58\ \frac{\text{ft}^3}{\text{lbm}}} = \boxed{3078\ \text{BTU/hr}}$$

Notice that h includes the energy used to vaporize the humidifying water. T_{water} is not relevant.

18. Walls: The common walls do not contribute to heat gain.

$$h_{\text{in}} = 1.65 \text{ BTU/hr-ft}^2\text{-°F} \quad [\text{film coefficient}]$$
$$h_{\text{out}} = 4.00 \text{ BTU/hr-ft}^2\text{-°F}$$
$$R_{\text{gypsum}} = 0.56 \text{ ft}^2\text{-°F-hr/BTU}$$
$$R_{\text{insulation}} = (3.5)(3.85) = 13.48 \text{ ft}^2\text{-°F-hr/BTU}$$
$$R_{\text{brick}} = 0.44 \text{ ft}^2\text{-°F-hr/BTU}$$
$$U = \frac{1}{\frac{1}{1.65} + \frac{1}{4.00} + 0.56 + 13.48 + 0.44} = 0.0652 \text{ BTU/hr-ft}^2\text{-°F}$$

Areas:

Walls:

$$(40)(4) + (5 + 10 + 5)(6) = 280 \text{ ft}^2$$

Windows:

$$(10 + 10)(6) = 120 \text{ ft}^2$$

Albuquerque data:

$$\text{latitude} = 35° \quad [\text{use } 32° \text{ or } 40° \text{ as necessary}]$$
$$F_{\text{shg}} = 31 \text{ BTU/hr-ft}^2 \quad [32° \text{ latitude table}]$$
$$\text{uncorrected } \Delta T_{\text{te}} = 17°\text{F}$$

Corrections:

- Daily swing is not given but may apply.
- $T_o - T_i = 96 - 75 = 21 \neq 20$; correction $= +1°\text{F}$

Corrected $\Delta T_{\text{te}} = 17 + 1 = 18°\text{F}$.

Psychrometric data:

$$h_i = 28.1 \text{ BTU/lbm}$$
$$h_o = 47.6 \text{ BTU/lbm}$$
$$v_o = 14.5 \text{ ft}^3\text{/lbm}$$

Ventilation:

$$(3)\left(25 \frac{\text{ft}^3}{\text{min}}\right)\left(60 \frac{\text{min}}{\text{hr}}\right) = 4500 \text{ ft}^3\text{/hr}$$

Heat gain:

For men-only occupants, scale the adjusted metabolic values by

$$\frac{390}{330} = 1.18$$

Latent:

Occupants:

$$(3)(105)(1.18) \approx 372 \text{ BTU/hr}$$

Sensible:

Occupants:

$$(3)(225)(1.18) \approx 797 \text{ BTU/hr}$$

Walls:

$$q = UA\Delta T_{\text{te}} = (0.0652)(280)(18) = 329 \text{ BTU/hr}$$

Windows: No data on internal shading is given.

Overhang has no effect on north-facing windows at this time of year. The actual analysis is beyond the scope of this solution.

$$q = A\left[C_s F_{\text{shg}} U(T_o - T_i)\right] = (120)\left[(0.93)(31) + (0.33)(96 - 75)\right] \approx 4291 \text{ BTU/hr}$$

Lights: For fluorescent lights,

$$q = (6)(4)(40 \text{ W})(3.413)(1.2) = 3931 \text{ BTU/hr}$$

The total sensible load is

$$q_s = 797 + 329 + 4291 + 3931 = 9348 \text{ BTU/hr}$$

(a) The total load on the air conditioner is

$$q = 372 + 9348 + \frac{\left(4500 \frac{\text{ft}^3}{\text{hr}}\right)\left(47.6 \frac{\text{BTU}}{\text{lbm}} - 28.1 \frac{\text{BTU}}{\text{lbm}}\right)}{14.5 \frac{\text{ft}^3}{\text{lbm}}}$$
$$= \boxed{15{,}772 \text{ BTU/hr}}$$

(b) At 42°F,

$$h = 10.05 \text{ BTU/lbm}$$
$$v = 0.01602 \text{ ft}^3\text{/lbm}$$

At 42°F + 15°F = 57°F,

$$h = 25.06 \text{ BTU/lbm}$$
$$m_{\text{water}} = \frac{q}{\Delta h} = \frac{15{,}772}{25.06 - 10.05} = 1051 \text{ lbm/hr}$$
$$Q = \frac{\left(1051 \frac{\text{lbm}}{\text{hr}}\right)\left(0.01602 \frac{\text{ft}^3}{\text{lbm}}\right)\left(7.48 \frac{\text{gal}}{\text{ft}^3}\right)}{60 \frac{\text{min}}{\text{hr}}} = \boxed{2.1 \text{ gpm}}$$

(c)
$$Q = \frac{R_{\text{air}} q_s}{T_i - T_{\text{in}}} = \frac{(55.3)(9348)}{(75 - 57)(60)} = \boxed{479 \text{ ft}^3\text{/min}}$$

MECHANICS OF MATERIALS

1. The bracket will rotate around the centroid of the weld group.

$\overline{x}$ is found by inspection:

$$\overline{x} = 2$$

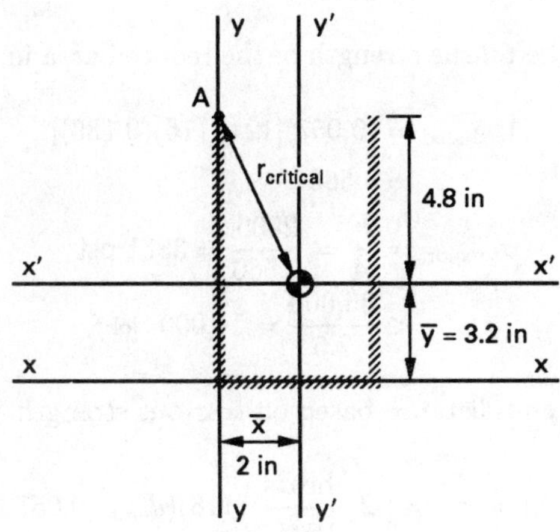

To find $\overline{y}$, consider the weld as a line.

$$\overline{y} = \frac{\sum \overline{y}_i L_i}{\sum L_i} = \frac{(4)(8) + (0)(4) + (4)(8)}{8 + 4 + 8}$$
$$= 3.2 \text{ in}$$

The critical weld point is found by inspection to be point A. At that point, vertical shear and torsional shear reinforce.

$$r_{\text{critical}} = \sqrt{(2)^2 + (4.8)^2}$$
$$= 5.2 \text{ in}$$

Considering the line to have a unity thickness, the polar moment of inertia is

$$J = I_{x'} + I_{y'}$$
$$I_{x'} = (1)(4)(3.2)^2 + 2\left[\frac{(1)(8)^3}{12} + (1)(8)(4 - 3.2)^2\right]$$
$$= 136.5 \text{ in}^4 \quad \left[\begin{array}{c}\text{units of in}^3 \text{ if unity} \\ \text{thickness is disregarded}\end{array}\right]$$
$$I_{y'} = 2\left[(1)(8)(2)^2\right] + \frac{(1)(4)^3}{12}$$
$$= 69.3 \text{ in}^4$$
$$J = 136.5 + 69.3 = 205.8 \text{ in}^4$$

However, the effective throat area is not 1, but

$$t_e = (0.707)\left(\frac{3}{8}\right) = 0.2651$$
$$J = (0.2651)(205.8) = 54.6 \text{ in}^4$$

The total weld throat area is

$$A = (0.2651)(8 + 4 + 8) = 5.3 \text{ in}^2$$

The force components are

$$P_x = P\cos 50° = 0.643P$$
$$P_y = P\sin 50° = 0.766P$$

The direct shear is P/A.

$$\tau_x = \frac{0.643P}{5.3} = 0.121P$$
$$\tau_y = \frac{0.766P}{5.3} = 0.145P$$

The eccentricity of the load is found as the distance from the line of action of the force to the weld group centroid.

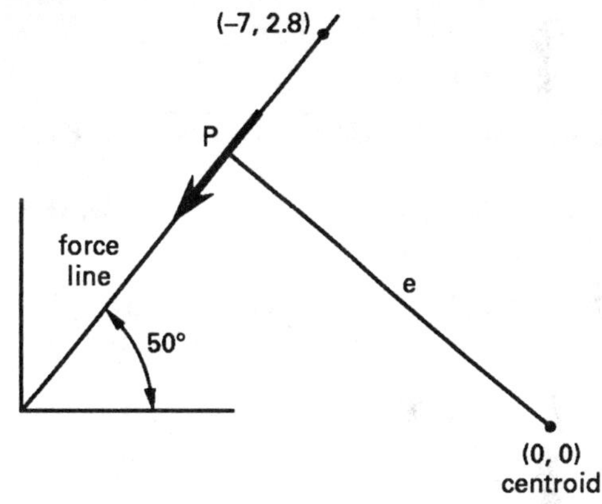

Force line:

$$\text{slope} = \tan 50° = 1.192$$
$$\text{known point} = (-7, 2.8)$$
$$y = mx + b$$
$$2.8 = (1.192)(-7) + b$$
$$b = 11.144$$

Putting this in general form,

$$1.192x - y + 11.144 = 0$$

The eccentricity is

$$e = \frac{|(1.192)(0) - (0) + 11.144|}{\sqrt{(1.192)^2 + (-1)^2}}$$
$$= 7.162 \text{ in}$$

The torsional moment is

$$T = Pe = 7.162P$$

At point A, the components of torsional shear are

$$\tau_x = \left(\frac{4.8}{5.2}\right)\tau = 0.923\tau$$
$$\tau_y = \left(\frac{2.0}{5.2}\right)\tau = 0.385\tau$$

The tortional shear is Tr/J.

$$\tau_x = \frac{(0.923)(7.162P)(5.2)}{54.6} = 0.630P$$
$$\tau_y = \frac{(0.385)(7.162P)(5.2)}{54.6} = 0.263P$$

Combining the direct and tortional shears,

$$\begin{aligned}\tau_{x,\text{total}} &= 0.121P + 0.630P \\ &= 0.751P \\ \tau_{y,\text{total}} &= 0.145P + 0.263P \\ &= 0.408P \\ \tau_{\text{maximum}} &= P\sqrt{(0.751)^2 + (0.408)^2} = 0.855P\end{aligned}$$

Nothing is known about the base metal, so it is presumed stronger than the rod.

For the E60 rod, 60 ksi is the ultimate strength.

Assume

$$S_{yt} \approx 0.75S_{ut}$$
$$S_{ys} = 0.577S_{yt}$$

Then,

$$0.855P = \frac{S_{ys}}{\text{FS}} = \frac{(0.577)(0.75)(60{,}000)}{2.5}$$

$$P = \boxed{12{,}150 \text{ lbf}}$$

2. From *Machinery's Handbook* or similar, a #29 drill has a diameter of 0.136 in.

The area per expanded rivet is

$$A = \frac{\pi}{4}d^2 = \left(\frac{\pi}{4}\right)(0.136)^2 = 0.01453 \text{ in}^2$$

Based on the rivet shear strength, the number of rivets needed to carry the load is

$$\frac{2000}{(0.01453)\left(\frac{22{,}000}{2.5}\right)} = 15.6 \quad \text{[say 16]}$$

Try 16 rivets in a single row. Each rivet carries 1/16 of the load.

Check the bearing stress.

$$\sigma_{\text{bearing}} = \frac{\frac{2000}{16}}{(0.136)(0.057)} = 16{,}125 \text{ psi}$$
$$< \frac{62{,}000}{2.5} = 24{,}800 \text{ psi} \quad \text{[ok]}$$

Check the tensile strength of the reduced area in shear.

$$\begin{aligned}A_{\text{reduced}} &= (0.057)[12 - (16)(0.136)] \\ &= 0.560 \\ \sigma_{\text{tension}} &= \frac{F}{A} = \frac{2000}{0.560} = 3571 \text{ psi} \\ &< \frac{35{,}000}{2.5} = 14{,}000 \quad \text{[ok]}\end{aligned}$$

Specify end-distance based on tear-out strength.

$$\text{area in shear} = \left(2\,\frac{\text{shears}}{\text{rivet}}\right)(16)(d_{\text{end}})(0.057 \text{ in})$$
$$= 1.824d_{\text{end}}$$

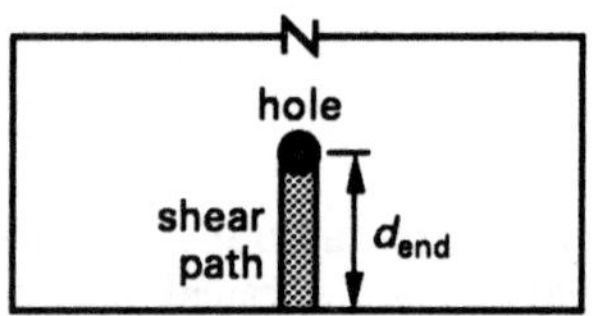

$$\tau = \frac{F}{A}$$
$$\frac{22{,}000}{2.5} = \frac{2000}{1.824d_{\text{end}}}$$
$$d_{\text{end}} = 0.125 \text{ in}$$

$$\boxed{\begin{array}{l} n = 16 \text{ rivets} \\ \text{one row} \\ \text{end distance} > 0.125 \text{ in}\end{array}}$$

3. From *Marks' Handbook*, manufacturers standard gage, #5 gage = 0.2092 in thickness.

(a)
$$I = (4 \text{ springs})\left[\frac{(2)(0.2092)^3}{12}\right]$$
$$= 0.006104 \text{ in}^4$$

Model a fixed-end beam free to translate at one end as two cantilevers.

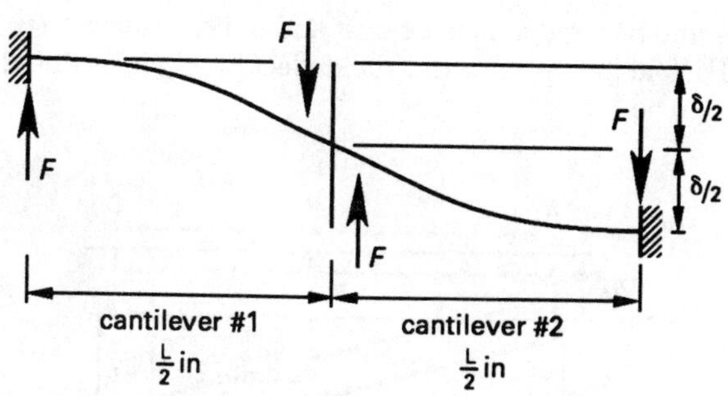

$$\delta = \frac{Pl^3}{3EI} = (2)\left[\frac{F\left(\frac{L}{2}\right)^3}{3EI}\right] = \frac{FL^3}{12EI}$$

Solving for F,

$$F = \frac{12EI\delta}{L^3} = \frac{(12)(2.9\times10^7)(0.006104)(2)}{(36)^3}$$

$$= \boxed{91.06 \text{ lbf}} \quad \left[\begin{array}{c}\text{total since } I \text{ was}\\ \text{for four springs}\end{array}\right]$$

(b) $M_{\text{maximum}} = F\left(\frac{L}{2}\right)$

$$\delta = \frac{Mc}{I} = \frac{(91.06)\left(\frac{36}{2}\right)\left(\frac{0.2092}{2}\right)}{0.006104}$$

$$= \boxed{28{,}090 \text{ psi}}$$

4. The loading direction is not known.

The relationship between the observed strain and and an assumed x axis is

$$\epsilon_{\text{observed}} = \frac{\epsilon_x+\epsilon_y}{2} + \frac{\epsilon_x-\epsilon_y}{2}\cos 2\theta + \frac{\epsilon_{xy}}{2}\sin 2\theta$$

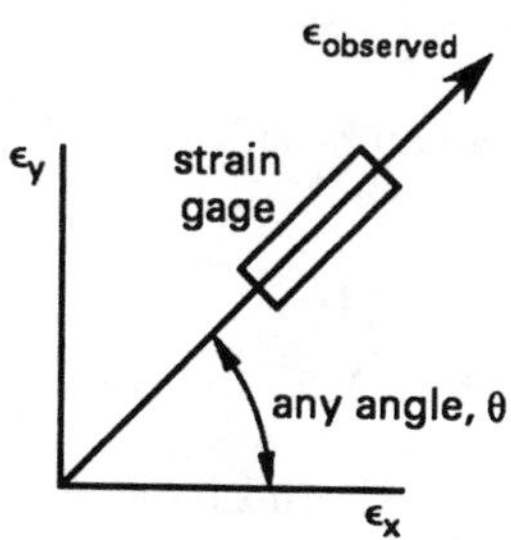

There are three unknowns: ϵ_x, ϵ_y, and ϵ_{xy}. Three equations must be used. If there is data for four (as in this case), the fourth can be used as a check.

Gage 1:

$$\theta_1 = 0$$
$$\cos 2\theta = 1$$
$$\sin 2\theta = 0$$
$$\epsilon_{\text{observed}} = \frac{\epsilon_x+\epsilon_y}{2} + \frac{\epsilon_x-\epsilon_y}{2} = \epsilon_x$$
$$\epsilon_x = 0.0005 \quad \text{[as expected]}$$

Gage 2:

$$\theta_2 = 45^\circ$$
$$\cos 2\theta = 0$$
$$\sin 2\theta = 1$$
$$\epsilon_{\text{observed}} = \frac{\epsilon_x+\epsilon_y}{2} + \frac{\epsilon_{xy}}{2}$$
$$\epsilon_x = 0.0005$$
$$-0.00245 = \frac{0.0005}{2} + \frac{\epsilon_y}{2} + \frac{\epsilon_{xy}}{2}$$
$$\epsilon_{xy} = -0.0054 - \epsilon_y$$

Gage 3:

$$\theta_3 = 90^\circ$$
$$\cos 2\theta = -1$$
$$\sin 2\theta = 0$$
$$\epsilon_{\text{observed}} = \frac{\epsilon_x+\epsilon_y}{2} - \frac{\epsilon_x-\epsilon_y}{2} = \epsilon_y$$
$$\epsilon_y = -0.0028 \quad \text{[as expected]}$$

Substituting the value of ϵ_y into the expression for ϵ_{xy} for gage 2,

$$\epsilon_{xy} = -0.0054 - (-0.0028)$$
$$= -0.0026$$

Gage 4 (use as a check):

$$\theta_4 = 135^\circ$$
$$\cos 2\theta = 0$$
$$\sin 2\theta = -1$$
$$\epsilon_{\text{observed}} = \frac{\epsilon_x+\epsilon_y}{2} - \frac{\epsilon_{xy}}{2}$$
$$(2)(0.00015) = 0.0005 - 0.0028 - (-0.0026)$$
$$0.0003 = 0.0003 \quad \text{[check]}$$

(a) Now that ϵ_x and ϵ_y are known, find the principal strains.

$$\epsilon_{\text{max,min}} = \frac{\epsilon_x+\epsilon_y}{2} \pm \tfrac{1}{2}\sqrt{(\epsilon_x-\epsilon_y)^2+\epsilon_{xy}^2}$$
$$= \frac{0.0005-0.0028}{2}$$
$$\pm \tfrac{1}{2}\sqrt{[0.0005-(-0.0028)]^2+(-0.0026)^2}$$
$$= -0.00115 \pm 0.0021$$
$$\epsilon_{\text{maximum}} = 0.00095$$
$$\epsilon_{\text{minimum}} = -0.00325$$

A strain in one direction is caused by stress in two perpendicular directions.

$$\sigma_{\text{maximum}} = \left(\frac{E}{1-\mu^2}\right)[\epsilon_{\text{maximum}} + \mu\epsilon_{\text{minimum}}]$$
$$= \left[\frac{300{,}000 \text{ MPa}}{1-(0.33)^2}\right] \times [0.00095 + (0.33)(-0.00325)]$$
$$= \boxed{-41.2 \text{ MPa}}$$

$$\sigma_{\text{minimum}} = \left(\frac{E}{1-\mu^2}\right)[\epsilon_{\text{minimum}} + \mu\epsilon_{\text{maximum}}]$$
$$= \left[\frac{300{,}000 \text{ MPa}}{1-(0.33)^2}\right] \times [-0.00325 + (0.33)(0.00095)]$$
$$= \boxed{-988.6 \text{ MPa}}$$

(b) $$\tau_{\text{maximum}} = \left[\frac{E}{2(1+\mu)}\right]\left[\tfrac{1}{2}\sqrt{(\epsilon_x - \epsilon_y)^2 + \tau_{xy}^2}\right]$$
$$= \left[\frac{300{,}000 \text{ MPa}}{(2)(1+0.33)}\right](0.0021)$$
$$= \boxed{236.8 \text{ MPa}}$$

5.

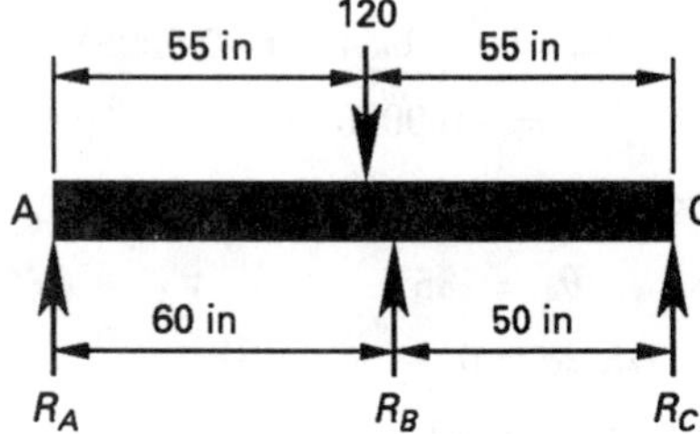

There are three unknown reactions. Normally, this would not be a difficult problem if the supports were rigid. The flexibility of the support affects the reaction.

First equation:

$$R_A + R_B + R_C = 120 \qquad \text{I}$$

Second equation:

Take moments about point A. Counter-clockwise moments are assumed to be positive.

$$\sum M_A = 0$$
$$110R_C + 60R_B - (55)(120) = 0$$
$$110R_C + 60R_B = 6600 \qquad \text{II}$$

Assume bar end C is least stiff and deflects more than A or B. The bar is rigid and the deflection changes linearly along the bar.

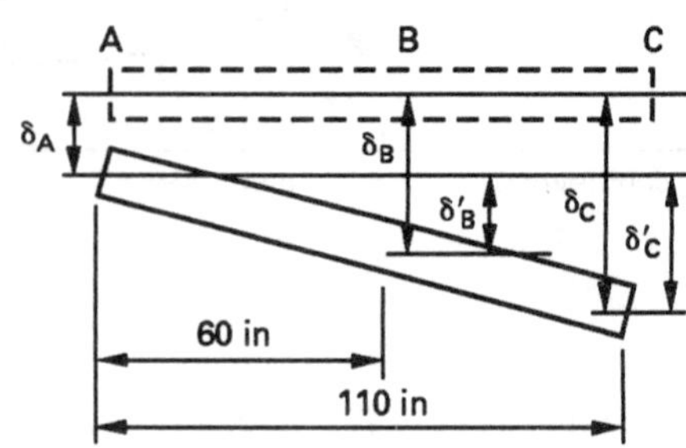

$$\delta'_C = \frac{110}{60}\delta'_B$$
$$\delta_C - \delta_A = \left(\frac{110}{60}\right)(\delta_B - \delta_A) = \frac{11}{6}\delta_B - \frac{11}{6}\delta_A$$

Simplifying further,

$$\frac{5}{6}\delta_A - \frac{11}{6}\delta_B - \delta_C = 0$$
$$5\delta_A - 11\delta_B - 6\delta_C = 0$$

This is the same relationship you get if end A is assumed to be the least stiff.

For a simply supported beam with a single concentrated load at midpoint,

$$\delta = \frac{FL^3}{48EI} \propto \frac{F}{I} \propto \frac{F}{(\text{side})^4}$$

(L and E are the same for all three supports.)

$$\delta_A \propto \frac{R_A}{(3.0)^4} = \frac{R_A}{81}$$
$$\delta_B \propto \frac{R_B}{(3.25)^4} = \frac{R_B}{111.6}$$
$$\delta_C \propto \frac{R_C}{(3.5)^4} = \frac{R_C}{150.1}$$

These expressions can be substituted into the deflection relationship.

$$\frac{5}{81}R_A - \frac{11}{111.6}R_B + \frac{6}{150.1}R_C = 0 \qquad \text{III}$$

Solving equations I, II, and III simultaneously yields

$$R_A = 41.313 \text{ kips}$$
$$R_B = 41.111 \text{ kips}$$
$$R_C = 37.576 \text{ kips}$$

The deflections on the support beams can be found.

$$\delta_A = \frac{FL^3}{48EI} = \frac{(41.313)(1000)(70)^3}{(48)(2.9\times10^7)\left(\frac{81}{12}\right)} = \boxed{1.508 \text{ in}}$$

(The 12 factor was included in the expression for I because $I = bh^3/12$.)

$$\delta_B = \frac{(41.111)(1000)(70)^3}{(48)(2.9\times10^7)\left(\frac{111.6}{12}\right)} = \boxed{1.089 \text{ in}}$$

$$\delta_C = \frac{(37.576)(1000)(70)^3}{(48)(2.9\times10^7)\left(\frac{150.1}{12}\right)} = \boxed{0.740 \text{ in}}$$

6.
$$I = \tfrac{1}{4}\pi r^4 = \tfrac{1}{4}\pi\left[\left(\frac{2.375}{2}\right)^4 - \left(\frac{2.067}{2}\right)^4\right] = 0.666 \text{ in}^4$$

$$A = \frac{\pi}{4}d^2 = \left(\frac{\pi}{4}\right)\left[(2.375)^2 - (2.067)^2\right] = 1.075 \text{ in}^2$$

The radius of gyration is

$$k = \sqrt{\frac{I}{A}} = \sqrt{\frac{0.666}{1.075}} = 0.787 \text{ in}$$

For a column with fixed ends, the effective length is

$$C = 0.65 \quad \left[\begin{array}{c}\text{ideally, } C = 0.5\text{, but no end can}\\ \text{truly be infinitely rigid}\end{array}\right]$$

$$L' = 0.65L$$

The slenderness ratio is

$$\frac{L'}{k} = \frac{(0.65)(11.5)(12)}{0.787} = 114$$

Check the minimum slenderness ratio for a long column with this material.

$$\left(\frac{L}{k}\right)_{\text{minimum}} = \sqrt{\frac{\pi^2 E}{S_y}} = \sqrt{\frac{\pi^2(2.9\times10^7)}{42{,}000}} = 82.6$$

Since $114 > 82.6$, this is a long column. Check the maximum compressive stress.

$$\sigma = \frac{F}{A} = \frac{12{,}000}{1.075} = 11{,}163 \text{ psi}$$

Since $11{,}163 < 42{,}000/2$, the Euler formula can be used.

The buckling load is

$$F_e = \frac{\pi^2 EI}{(L')^2} = \frac{\pi^2(2.9\times10^7)(0.666)}{[(0.65)(11.5)(12)]^2} = 23{,}691 \text{ lbf}$$

Since $12{,}000 < 23{,}691$ there is $\boxed{\text{no buckling}}$

The factor of safety against buckling is

$$\text{FS}_{\text{buckling}} = \frac{23{,}691}{12{,}000} = \boxed{1.97}$$

7. (a) The beam is neither simply supported nor fixed at its ends. As the beam deflects, the ends will rotate through an angle, θ_0. From compatibility, the torsion bars will also twist an angle, θ_0. The bars impart a resisting moment, M_0, to the beam ends. The freebody diagram is

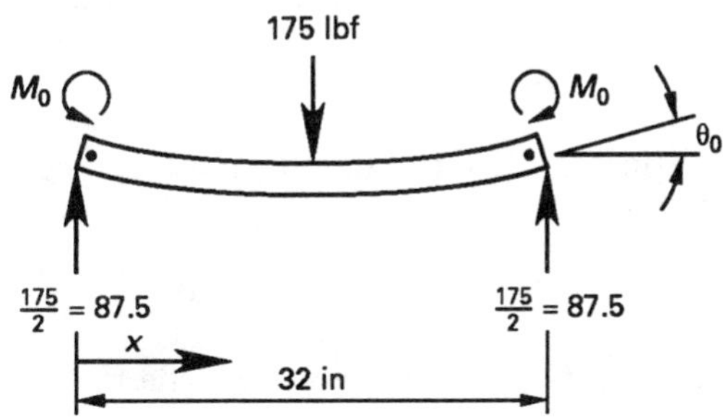

The shear diagram is

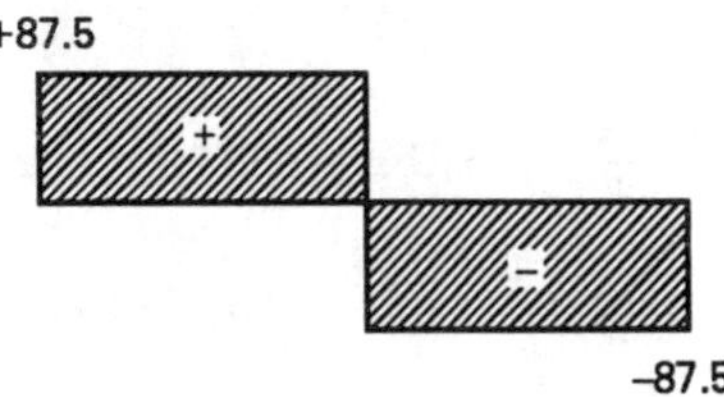

The moment equation and moment diagram can be derived from the shear diagram. Since the beam is symmetrical, only half the beam is needed.

$$V(x) = 87.5 \text{ lbf} \quad [0 \le x \le 16 \text{ in}]$$

$$M(x) = \int V(x)dx = \int 87.5dx = 87.5x + M_0 \qquad \text{I}$$

The moment diagram can now be drawn. M_0 is unknown. It must be negative, however, because it is contrary to the moment induced by the 175 lbf force.

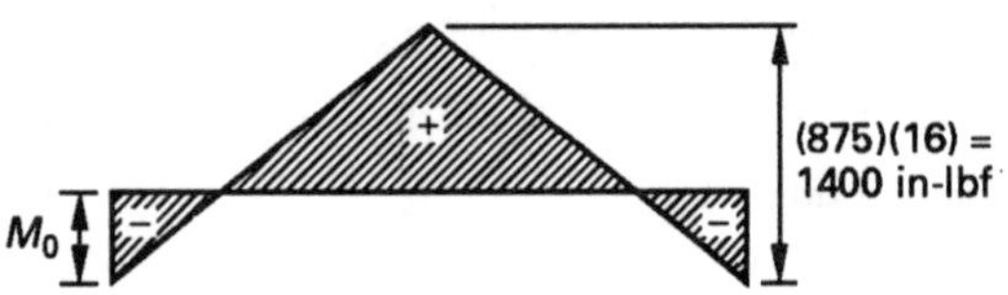

Another equation for M_0 can be derived from the twist of the torsion bars,

$$\theta_0 = \frac{TL}{JG} = \frac{\left(\frac{M_0}{2}\right)L}{JG}$$

For each bar,

$$J = \frac{\pi d^4}{32} = \frac{\pi(1.125)^4}{32} = 0.15726 \text{ in}^4$$
$$G = 1.1 \times 10^7 \text{ psi} \quad \text{[given]}$$
$$L = 4.125 \text{ in}$$

Since there are two bars at each beam end, the applied moment is

$$M_0 = 2T = \frac{2JG\theta_0}{L} = \frac{(2)(0.15726)(1.1 \times 10^7)\theta_0}{4.125}$$
$$= 838{,}720\theta_0 \qquad \text{II}$$

Substituting Eq. II into Eq. I (θ_0 must be negative to keep M_0 negative),

$$M(x) = 87.5x + 838{,}720\theta_0 \qquad \text{III}$$

Integrate Eq. III to get the angular deflection.

Note: $\dfrac{d^2y}{dx^2} = \dfrac{M(x)}{EI}$

$$\frac{dy}{dx} = \frac{1}{EI}\int M(x)dx$$
$$= \frac{1}{EI}\int (87.5x + 838{,}720\theta_0)dx$$
$$= \frac{1}{EI}\int \frac{87.5x^2}{2} + 838{,}720\theta_0 x + C_2 \qquad \text{IV}$$

For small angles,

$$\theta = \tan\theta = \frac{dy}{dx} \quad [\theta \text{ in radians}]$$

The two boundary conditions are

(A) $\theta = \theta_0$ at $x = 0$

(B) $\theta = 0$ at $x = \dfrac{L}{2} = 16$ in

From (A),

$$\theta_0 = \left(\frac{1}{EI}\right)(0 + 0 + C_2)$$
$$C_2 = EI\theta_0 = (3 \times 10^7)\left[\frac{(2.25)(3.25)^3}{12}\right]\theta_0$$
$$= 1.93096 \times 10^8\theta_0$$

From (B),

$$0 = \left(\frac{1}{1.93096 \times 10^8}\right) \times \left[\frac{(87.5)(16)^2}{2} + (838{,}720\theta_0)(16) + 1.93096 \times 10^8\theta_0\right]$$
$$0 = 5.8 \times 10^{-5} + 0.069497\theta_0 + \theta_0$$
$$\theta_0 = -5.4232 \times 10^{-5} \text{ radians}$$

Now that θ_0 is known, C_2 can be found.

$$C_2 = 1.93096 \times 10^8\theta_0$$
$$= (1.93096 \times 10^8)(-5.4232 \times 10^{-5})$$
$$= -10{,}472$$

The end moment is

$$M_0 = 838{,}720\theta_0 = (838{,}720)(-5.4232 \times 10^{-5})$$
$$= -45.486 \text{ in-lbf}$$

The torsion in each bar is

$$T = \frac{M_0}{2} = \frac{-45.486}{2} = \boxed{-22.743 \text{ in-lbf}}$$

(b) From Eq. I,

$$M(x) = 87.5x + M_0 = 87.5 - 45.486$$

At $x = 16$,

$$m_{\text{maximum}} = (87.5)(16) - 45.486 = \boxed{1354.5 \text{ in-lbf}}$$

Check:

$$1354.5 + 45.486 = 1400 \quad \text{[ok]}$$

(c) The total deflection is the sum of the beam and bar deflections.

Beam deflection:

From Eq. IV,

$$\frac{dy}{dx} = \left(\frac{1}{1.93096 \times 10^8}\right)\left[\frac{87.5x^2}{2} + (838{,}720)(-5.4232 \times 10^{-5})x - 10{,}472\right]$$

$$= \left(\frac{1}{1.93096 \times 10^8}\right)\left(\frac{87.5x^2}{2} - 45.485x - 10{,}472\right)$$

$$y = \int \frac{dy}{dx}$$

$$= \left(\frac{1}{1.93096 \times 10^8}\right) \times \left(\frac{87.5x^3}{6} - \frac{45.485x^2}{2} - 10{,}472x + C_3\right)$$

$y = 0$ at $x = 0$, so $C_3 = 0$.

$$y_{\text{maximum}} = y(16) = \frac{\frac{(87.5)(16)^3}{6} - \frac{(45.485)(16)^2}{2} - (10{,}472)(16)}{1.93096 \times 10^8} = -0.0005885 \text{ in}$$

Bar deflection:

Each bar is a fixed-end beam with ends free to translate.

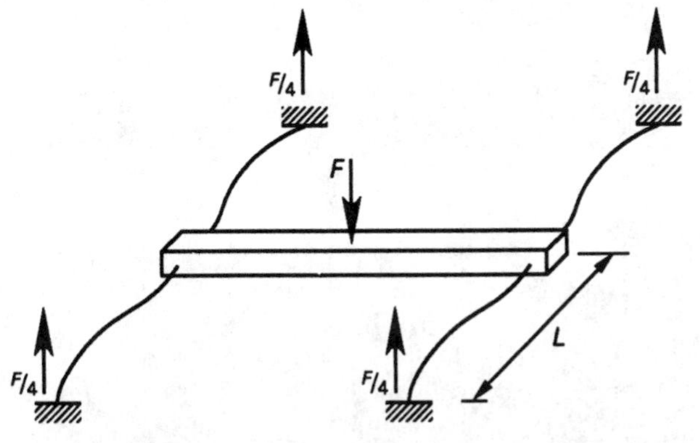

Think of each torsion bar as being two cantilevers of length $l = L/2$, each supporting $F/4$.

$$y = \frac{Pl^3}{3EI} = (2)\left[\frac{\left(\frac{F}{4}\right)\left(\frac{L}{2}\right)^3}{3EI}\right] = \frac{\left(\frac{F}{4}\right)L^3}{12EI}$$

$$I_{\text{bars}} = \frac{\pi D^4}{64} = \frac{\pi(1.125)^4}{64} = 0.07863 \text{ in}^4$$

$$y_{\text{maximum}} = \frac{\left(\frac{174}{4}\right)(4.125)^3}{(12)(3 \times 10^7)(0.07863)} = 0.000108 \text{ in}$$

The total deflection is

$$y_{\text{total}} = y_{\text{beam}} + y_{\text{bars}} = 0.0005885 + 0.000108 = \boxed{0.0006965 \text{ in}}$$

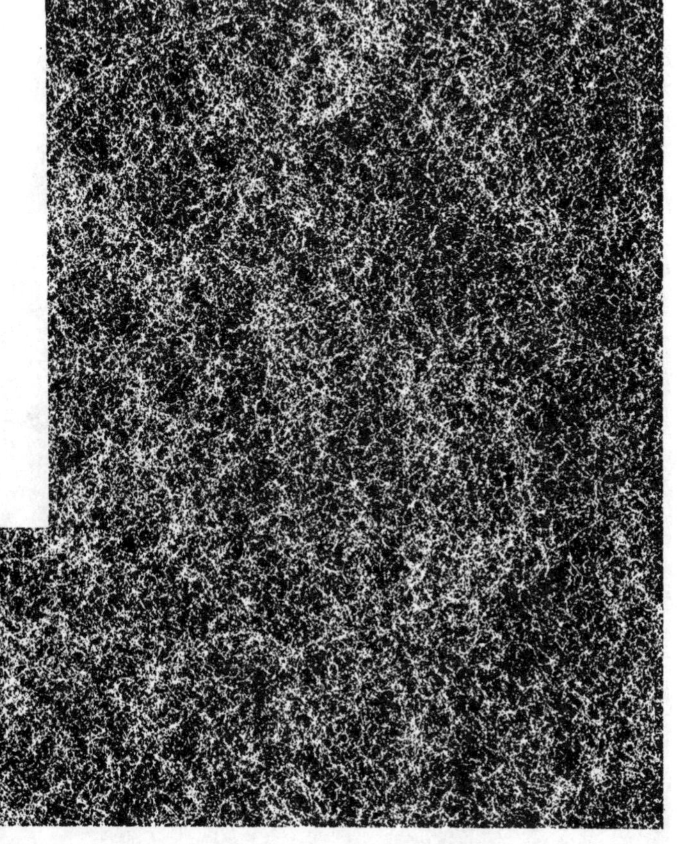

FAILURE THEORY

1.
$$I_{\text{original}} = \frac{(0.28)(1.4)^3}{12} = 0.064 \text{ in}^4$$
$$I_{\text{damaged}} = \frac{(0.28)(1.05)^3}{12} = 0.027 \text{ in}^4$$

For an infinite life, the allowable stress is $S_e = 40{,}000$ psi.

The actual stress (reduced by 200%) is

$$\frac{40{,}000}{2} = 20{,}000 \text{ psi}$$

The handle was originally designed for a bending stress.

$$M_{\text{original}} = \frac{\sigma I_{\text{original}}}{c_{\text{original}}} = \frac{(20{,}000)(0.064)}{\frac{1.4}{2}} = 1829 \text{ in-lbf}$$

The new stress will be

$$\sigma_{\text{damaged}} = \frac{Mc_{\text{damaged}}}{I_{\text{damaged}}} = \frac{(1829)\left(\frac{1.05}{2}\right)}{0.027} = 35{,}564 \text{ psi}$$

Assume the bar has been heated and cooled so that the annealed conditions apply. Disregard the previous history of the part, notch sensitivity, stress concentration factors, surface finish, etc., all of which could easily double the localized stress.

$$\sigma_{\text{damaged}} > S_{e,\text{annealed}}$$

Failure will occur.

Draw an approximate S-N diagram.

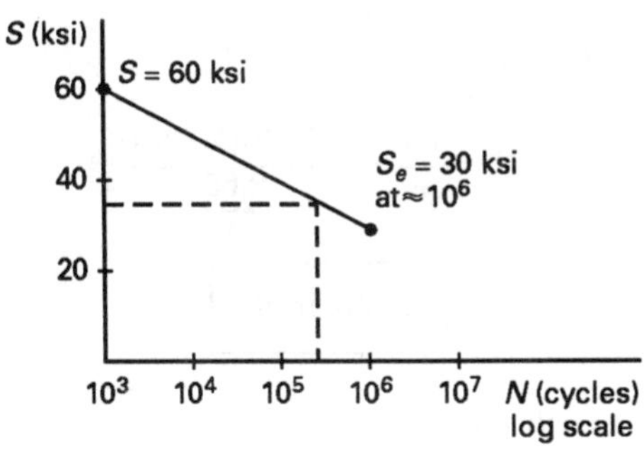

Failure will occur between 10^5 and 10^6 cycles.

2. Assume:

$$\text{bolt shank length} = \text{grip of parts}$$
$$E_{\text{rod cap}} = E_{\text{bolts}}$$

(No data is given for either of these.)

(a) The stress area for the bolt is

$$A = 0.0876 \text{ in}^2$$

The spring constant is

$$k_{\text{bolt}} = \frac{0.0876E}{L}$$
$$k_{\text{rod cap}} = \frac{0.32E}{L}$$

Let $k_{\text{bolt}} = 1$. Then, since E and L values are the same,

$$k_{\text{rod cap}} = \frac{A_{\text{rod cap}}}{A_{\text{bolt}}} = \left(\frac{0.32}{0.0876}\right)(1) = 3.65$$

The initial preloading is

$$\sigma_i = (0.80)(80{,}000) = 64{,}000 \text{ psi}$$

With compression negative, the bolt stress is

$$\sigma_{\text{bolt,maximum}} = 64{,}000 + \frac{(1)\left(\frac{7000}{2}\right)}{(1+3.65)(0.0876)} = 72{,}592 \text{ psi [tension]}$$

$$\sigma_{\text{bolt,minimum}} = 64{,}000 + \frac{(1)\left(\frac{-24{,}000}{2}\right)}{(1+3.65)(0.0876)} = 34{,}541 \text{ psi [tension]}$$

The stress concentration factor is 2.65.

$$\sigma_{\text{alt}} = \left(\tfrac{1}{2}\right)(\sigma_{\text{maximum}} - \sigma_{\text{minimum}}) = \left(\tfrac{1}{2}\right)(72{,}592 - 34{,}541)(2.65) = \boxed{50{,}418 \text{ psi}}$$

(b) Stress is alternating, so use a Goodman diagram.

$$\sigma_{\text{alt}} = 50{,}418$$
$$\sigma_{\text{mean}} = \left(\tfrac{1}{2}\right)(72{,}592 + 34{,}541) = 53{,}567$$

The stress concentration factor is not applied to σ_{mean}.

Since $\sigma_{\text{alt}} > S_e$, **the design is not acceptable**

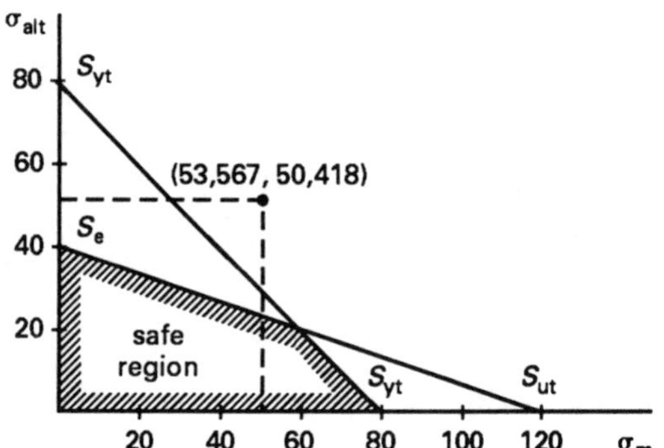

(c) The following factors are greatly different.

- modulus of elasticity
- durability and endurance limit
- expected life
- weight
- size to obtain equivalent strength
- material cost
- manufacturing method
- thermal expansion
- galvanic action and corrosion

3. Check thinwall stress assumptions.

$$\frac{t}{d} = \frac{0.20}{2.25} \approx 0.09 < \frac{1}{10} \quad \text{[ok]}$$

The hoop stress is

$$\sigma_h = \frac{pr}{t} = \frac{(750)\left(\frac{2.25}{2}\right)}{0.20} = 4219 \text{ psi} \quad \text{[tension]}$$

The torsional stress is unknown.

The combined stress is

$$\sigma_1, \sigma_2 = \left(\tfrac{1}{2}\right)(4219) \pm \left(\tfrac{1}{2}\right)\sqrt{(4219)^2 + (2\tau)^2}$$

Try #1:

Assume $\tau = 5000$ psi.

$$\sigma_1, \sigma_2 = 7536, -3317 \text{ psi}$$

Draw the maximum shear stress envelope.

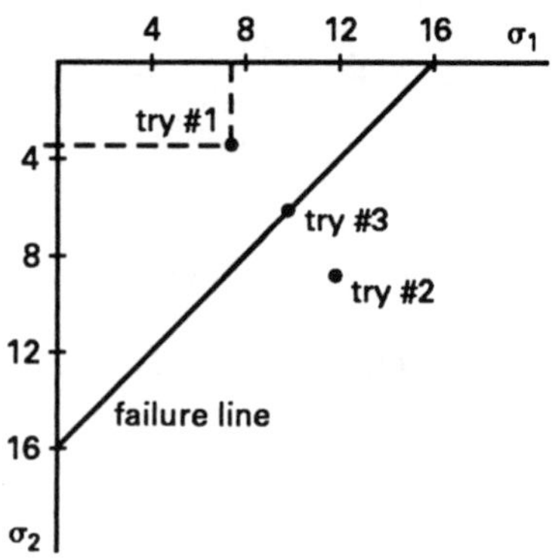

Try #2:

Assume $\tau = 10{,}000$ psi.

$$\sigma_1, \sigma_2 = 12{,}330, -8111 \text{ psi}$$

This is outside the failure envelope.

Try #3:

Assume $\tau = 7800$ psi.

$$\sigma_1, \sigma_2 \approx 10{,}190, -5970 \text{ psi}$$

This is approximately right on the failure line. (Note $\sigma_1, \sigma_2 \approx 16{,}000$.)

The outside diameter is

$$\begin{aligned} d_{\text{outside}} &= d_1 + 2t = 2.25 + (2)(0.20) \\ &= 2.65 \end{aligned}$$

The torsional moment of inertia is

$$\begin{aligned} J &= \left(\frac{\pi}{32}\right)\left(d_{\text{outside}}^4 - d_i^4\right) \\ &= \left(\frac{\pi}{32}\right)\left[(2.65)^4 - (2.25)^4\right] \\ &= 2.325 \text{ in}^4 \end{aligned}$$

Since $\tau = Tr/J$,

$$T = \frac{\tau J}{r} = \frac{(7800)(2.325)}{\frac{2.65}{2}} = \boxed{13{,}687 \text{ in-lbf}}$$

Alternate solution:

Try #1:

$$\tau = 5000 \text{ psi}$$

$$\sigma_1, \sigma_2 = 7536, -3317 \text{ psi}$$

Von Mises stress is

$$\sigma' = \sqrt{(7536)^2 + (-3317)^2 - (7536)(-3317)}$$
$$= 9633 \text{ psi}$$
$$\text{FS} = \frac{16{,}000}{9633} = 1.66$$

Try #2:

$$\tau = 10{,}000 \text{ psi}$$
$$\sigma_1, \sigma_2 = 12{,}330, -8111 \text{ psi}$$
$$\sigma' = 17{,}828 \text{ psi}$$
$$\text{FS} = \frac{16{,}000}{17{,}828} = 0.90$$

Try #3:

$$\tau = 5000 + (10{,}000 - 5000)\left(\frac{1.66 - 1.00}{1.66 - 0.9}\right)$$
$$= 9342 \text{ psi} \quad \text{[say 9000]}$$
$$\tau = 9000 \text{ psi}$$
$$\sigma_1, \sigma_2 = 11{,}354, -7134 \text{ psi}$$
$$\sigma' = 16{,}150 \text{ psi}$$
$$\text{FS} = 0.99$$

Try #4:

$$\tau = 8900 \text{ psi}$$
$$\sigma_1, \sigma_2 = 11{,}257, -7037 \text{ psi}$$
$$\sigma' = 15{,}983 \text{ psi}$$
$$\text{FS} = 1.00$$

Finishing,

$$d_{\text{outside}} = 2.65 \text{ in}$$
$$J = 2.325 \text{ in}^4$$
$$T = \frac{\tau J}{r} = \frac{(8900)(2.325)}{\frac{2.65}{2}}$$
$$= \boxed{15{,}617 \text{ in-lbf}}$$

4. (a) $$J = \frac{TL}{G\phi}$$

T/ϕ is the torsional spring constant in in-lbf/radian.

$$J = \frac{(3600)(32)}{11.5 \times 10^6} = 0.010 \text{ in}^4$$
$$d = \sqrt[4]{\frac{32J}{\pi}} = \sqrt[4]{\frac{(32)(0.010)}{\pi}}$$
$$= \boxed{0.565 \text{ in}}$$

It might be desirable to select a standard size, but this would change performance.

(b) $$S_{ys} = S_{yt} = (0.577)(60{,}000)$$
$$= \boxed{34{,}620 \text{ psi}}$$

(c) Since $\tau = Tr/J$, using FS = 1.0,

$$T = \frac{(34{,}620)(0.010)}{\frac{0.565}{2}} = \boxed{1225 \text{ in-lbf}}$$

(d) $$8° = \left(\frac{8}{360}\right)(2\pi) = 0.1396 \text{ radians}$$
$$\phi = \frac{TL}{GJ} \text{ and } \tau = \frac{Tr}{J} \quad [T \text{ is common}]$$
$$\tau = \frac{\left(\frac{\phi GJ}{L}\right)r}{J} = \frac{\phi G r}{L}$$
$$= \frac{(0.1396)(11.5 \times 10^6)\left(\frac{0.565}{2}\right)}{32}$$
$$= \boxed{14{,}173 \text{ psi}}$$

5. Check the thinwall assumption.

$$\frac{t}{d} = \frac{0.25}{(2)(3.5)} = 0.036 < \frac{1}{10} \quad \text{[ok]}$$

The hoop stress is

$$\sigma_h = \frac{pr}{t} = \frac{(800)(3.5)}{0.25} = 11{,}200 \text{ psi}$$
$$r_o = r_m + \frac{t}{2} = 3.5 + \frac{0.25}{2} = 3.625 \text{ in}$$
$$r_i = 3.5 - \frac{0.25}{2} = 3.375 \text{ in}$$
$$J = \left(\frac{\pi}{2}\right)\left[(3.625)^4 - (3.375)^4\right] = 67.43 \text{ in}^4$$

The shear stress is

$$\tau = \frac{Tr}{J} = \frac{(190{,}000)(3.625)}{67.43} = 10{,}214 \text{ psi}$$
$$\tau_{\text{maximum}} = \left(\tfrac{1}{2}\right)\sqrt{(11{,}200)^2 + [(2)(10{,}214)]^2}$$
$$= 11{,}648 \text{ psi}$$
$$\sigma_1, \sigma_2 = \left(\tfrac{1}{2}\right)(11{,}200) \pm 11{,}648$$
$$= 17{,}248, -6048 \text{ psi}$$

Draw the maximum shear stress failure envelope. The σ_1 and σ_2 intercepts can be found graphically as approximately 23,000 psi, or exactly as

$$17{,}248 + 6048 = 23{,}296 \text{ psi}$$

The factor of safety based on the equivalent stress is

$$\text{FS} = \frac{50{,}000}{\sigma'} = \frac{50{,}000}{23{,}296} = \boxed{2.15}$$

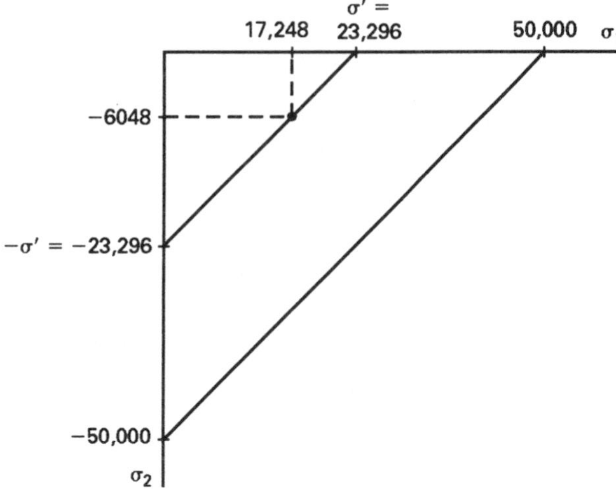

Alternate solution:

If the Von Mises theory is used,

$$\begin{aligned}(\sigma')^2 &= \sigma_1^2 + \sigma_2^2 - \sigma_1\sigma_2 \\ &= (17{,}248)^2 + (-6048)^2 - (17{,}248)(-6048) \\ \sigma' &= 20{,}938 \text{ psi} \\ \text{FS} &= \frac{S_{yt}}{\sigma'} = \frac{50{,}000}{20{,}938} \\ &= \boxed{2.39}\end{aligned}$$

Note that a factor of safety calculated as $\frac{S_{ys}}{\tau} = \frac{0.577S_{yt}}{\tau}$ does not yield the same answer, and is incorrect.

6.

$$I = \frac{(1.5)(1.5)^3}{12} = 0.422 \text{ in}^4$$

$$M_{\text{maximum}} = \frac{FL}{8} = \frac{(1500)(48)}{8} = 9000 \text{ in-lbf}$$

This is a rectangular bar in bending. A stress concentration factor is needed.

From Shigley's *Mechanical Engineering Design*, with

$$\frac{D}{d} = \frac{4.5}{1.5} = 3.0$$

$$\frac{r}{d} = \frac{0.25}{1.5} = 0.167$$

The stress concentration factor is $k = 1.6$.

The design stress is

$$\begin{aligned}\sigma &= k(\text{FS})\left(\frac{Mc}{I}\right) \\ &= (1.6)(1.5)\left[\frac{(9000)\left(\frac{1.5}{2}\right)}{0.422}\right] \\ &= 38{,}389 \text{ psi}\end{aligned}$$

This must coincide with S_e, so

$$\begin{aligned}S_{ut} &= \frac{S_e}{0.33} = \frac{38{,}389}{0.33} \\ &= \boxed{116{,}330 \text{ psi}}\end{aligned}$$

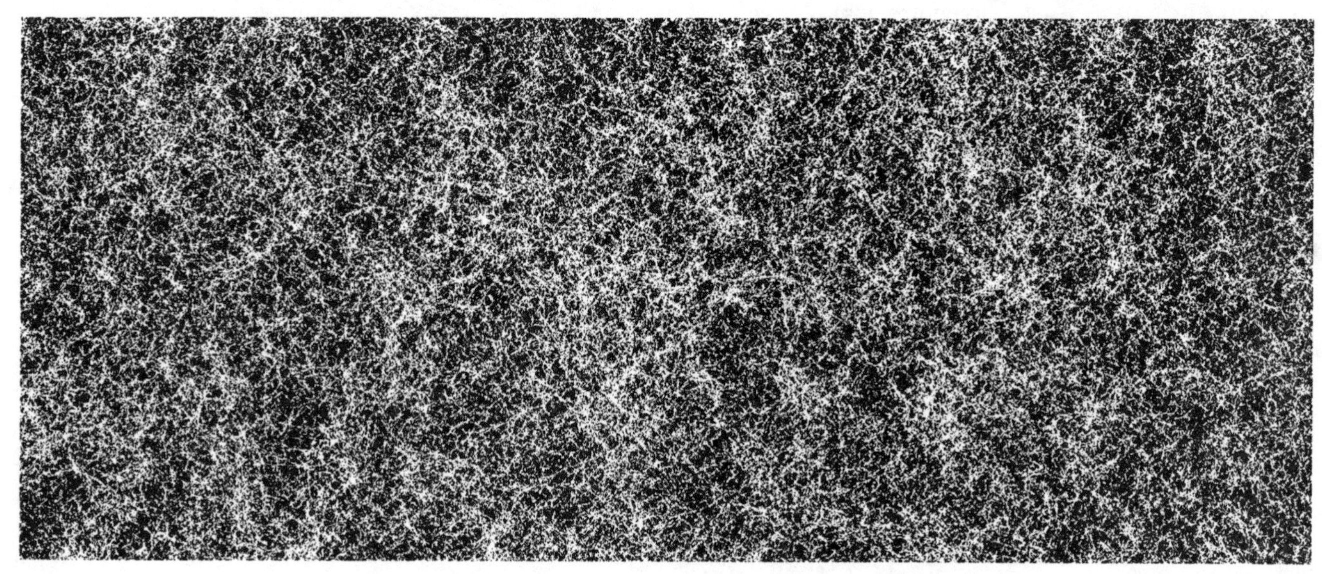

MACHINE DESIGN

1. Neglect radial deformation of the shaft during rotation.

For aluminum,

$$E = 10 \times 10^6 \text{ psi}$$
$$\mu = 0.33$$
$$\rho = 173 \text{ lbm/ft}^3 = 0.100 \text{ lbm/in}^3$$

For steel,

$$E = 2.9 \times 10^7 \text{ psi}$$
$$\mu = 0.30$$

The tangential stress is

$$\sigma_t = E\epsilon_t = E\left(\frac{\Delta \text{ circumference}}{\text{circumference}}\right) = E\left(\frac{2\pi\Delta r}{2\pi r}\right)$$

Solving for Δr,

$$\Delta r = \frac{r\sigma_t}{E}$$

Calculate σ_t ($g = 386.4$ in/sec^2).

$$\Delta r = \left(\tfrac{1}{4}\right)\left(\frac{\rho\omega^2 r_i}{386.4E}\right)\left[(3+\mu)r_o^2 + (1-\mu)r_i^2\right]$$
$$\omega = 2\pi f = 2\pi\left(\frac{\text{rpm}}{60}\right)$$
$$\frac{0.0045}{2} = \left[\frac{(0.100)(2\pi)^2(\text{rpm})^2(1)}{(4)(386.4)(60)^2(10 \times 10^6)}\right] \times \left[(3.33)(6)^2 + (0.67)(1)^2\right]$$
$$\text{rpm} = \boxed{16{,}219}$$

2.

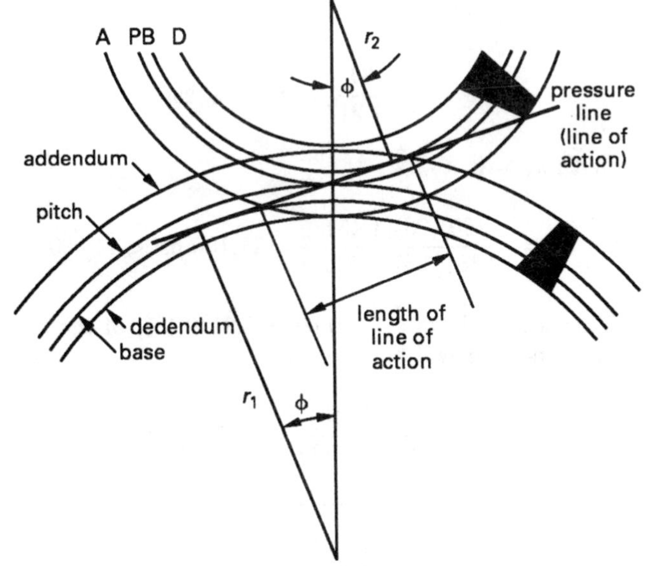

$$r_1 + r_2 = 8.5 \text{ in}$$
$$\frac{r_1}{r_2} = 2.5$$
$$r_1 = 2.5r_2$$
$$2.5r_2 + r_2 = 8.5 \text{ in}$$
$$r_2 = 2.43 \text{ in}$$
$$r_1 = 8.5 - 2.43 = 6.07$$
$$d_1 = (2)(6.07) = \boxed{12.14 \text{ in}}$$
$$d_2 = (2)(2.43) = \boxed{4.86 \text{ in}}$$

The contact ratio is

$$\text{CR} = \frac{L}{\left(\frac{\pi}{P}\right)\cos\phi}$$

The line of action, L, is tangent to both base circles and can be found graphically. The length of the line of action is the distance between intersections of the line of action and addendum circles.

L can be calculated as

$$L = \sqrt{\left(r_1 + \frac{m_1}{P}\right)^2 - r_1^2\cos^2\phi} - r_1\sin\phi + \sqrt{\left(r_2 + \frac{m_2}{P}\right)^2 - r_2^2\cos^2\phi} - r_2\sin\phi$$

For full-depth teeth, $m = 1$. For stub teeth, $m = 0.8$.

The length of the line of action is

$$L = \sqrt{\left(6.07 + \frac{1}{P}\right)^2 - (6.07)^2(\cos 20^\circ)^2} - (6.07)(\sin 20^\circ) + \sqrt{\left(2.43 + \frac{1}{P}\right)^2 - (2.43)^2(\cos 20^\circ)^2} - (2.43)(\sin 20^\circ)$$
$$= \sqrt{\left(6.07 + \frac{1}{P}\right)^2 - 32.535} + \sqrt{\left(2.43 + \frac{1}{P}\right)^2 - 5.214} - 2.907$$

By trial and error,

P	L	CR
1.0	3.83	1.30
1.25	3.21	1.35
2.0	2.19	1.48
2.5	1.82	1.54

$$P = \boxed{2.5}$$

Now the numbers of teeth can be found.

$$n = Pd$$
$$n_1 = (2.5)(12.14) = 30.35$$
$$n_2 = (2.5)(4.86) = 12.15$$

Say

$$n_1 = \boxed{30}$$
$$n_2 = \boxed{12}$$

3. The required pressure at 190°F is

$$p = \frac{T}{2\pi r_{\text{shaft}}^2 Lf} = \frac{(3000)(12)}{2\pi\left(\frac{6}{2}\right)^2(8)(0.16)}$$
$$= 497.4 \text{ psi}$$

At this point, in a "real" design, it would be appropriate to add a factor of safety (e.g., FS = 2) and a stress concentration factor ($k \approx 4$) since the coupling is shorter than the shaft. These are mentioned, but not implemented.

Shaft, external pressure at 190°F:

Using $r_o = 3.0$ and $r_i = 0$,

$$\sigma_{ro} = -p = -497.4 \text{ psi}$$
$$\sigma_{co} = \frac{-\left[(3)^2 + (0)^2\right](497.4)}{(3)^2 - (0)^2} = -497.4 \text{ psi}$$

With $\sigma_L = 0$,

$$\Delta D_{\text{shaft}} = \frac{(6)[-497.4 - (0.30)(-497.4)]}{2.9 \times 10^7}$$
$$= -7.204 \times 10^{-5} \text{ in}$$

Coupling, internal pressure at 190°F:

Using $r_o = 3.5$ and $r_i = 3.0$,

$$\sigma_{ri} = -497.4 \text{ psi}$$
$$\sigma_{ci} = \frac{\left[(3.5)^2 + (3.0)^2\right](497.4)}{(3.5)^2 - (3.0)^2} = 3252.2 \text{ psi}$$

With $\sigma_L = 0$,

$$\Delta D_{\text{coupling}} = \frac{(6)[3252.2 - (0.33)(-497.4)]}{1.6 \times 10^7}$$
$$= 1.281 \times 10^{-3} \text{ in}$$

If unconstrained, the shaft diameter at 190°F would be

$$D_{\text{shaft},190^\circ\text{F}} = D_o + \Delta D_{\text{shaft}} \approx 6.000 + 7.204 \times 10^{-5} \text{ in}$$

The unconstrained diameter at 70°F is

$$D_{\text{shaft},70^\circ\text{F}} = D_{\text{shaft},190^\circ\text{F}}(1 + \alpha\Delta T)$$
$$= (6.000 + 7.204 \times 10^{-5} \text{ in}) \times \left[1 + (6.5 \times 10^{-6})(70 - 190)\right]$$
$$= (6.00007204)(1 - 0.00078)$$
$$= 5.995392 \text{ in}$$

If unconstrained, the coupling diameter at 190°F would be

$$D_{\text{coupling},190^\circ\text{F}} = D_o + \Delta D_{\text{coupling}}$$
$$\approx 6.000 - 1.281 \times 10^{-3} \text{ in}$$

The unconstrained diameter at 70°F is

$$D_{\text{coupling},70^\circ\text{F}} = D_{\text{coupling},190^\circ\text{F}}(1 + \alpha\Delta T)$$
$$= (6.000 - 1.281 \times 10^{-3} \text{ in}) \times \left[1 + (10.2 \times 10^{-6})(70 - 190)\right]$$
$$= (5.998719)(1 - 0.001224)$$
$$= 5.991377$$

The diametral interference at 70°F is

$$D_{\text{shaft},70^\circ\text{F}} - D_{\text{coupling},70^\circ\text{F}} = 5.995392 - 5.991377$$
$$= \boxed{0.004015 \text{ in}}$$

4. Since speed and temperature are low, fatigue and durability are not considered.

Use the Lewis beam strength theory.

With 34 teeth,

$$Y = 0.371$$

The diametral pitch is

$$P = \frac{34}{1.21} = 28.1 \quad \text{[say 28]}$$

With $k_d = 1$,

$$\sigma = \frac{(60)(28)(1.45)}{(0.35)(0.371)(1.69)} = 11{,}100 \text{ psi} < 14{,}000$$

This seems ok, but 1 − 0.69 = 0.31 (31%) of the time only one gear will be in contact.

$$\sigma = \frac{(60)(28)(1.45)}{(0.35)(0.371)(1)} = 18{,}760 > 14{,}000$$

$$\boxed{\text{no good}}$$

5. The stress area of the bolt is

$$A_{\text{bolt}} = 0.6624 \text{ in}^2$$

The nominal bolt area is

$$A = \left(\frac{\pi}{4}\right)(1) = 0.7854 \text{ in}^2$$

Spring constants:

Disregard the rigid support.

$$\begin{aligned} k_{\text{parts}} &= \frac{AE}{L} = \frac{(1.4)(2.9 \times 10^7)}{1.50 + 0.50} \\ &= 2.03 \times 10^7 \text{ lbf/in} \end{aligned}$$

With no threads in the spacer, the nominal bolt area is used to calculate k_{bolt}.

$$\begin{aligned} k_{\text{bolt}} &= \frac{(0.7854)(2.9 \times 10^7)}{0.75 + 1.50 + 0.50} \\ &= 8.28 \times 10^6 \text{ lbf/in} \end{aligned}$$

Force in bolt:

$$\begin{aligned} F_{\text{bolt,min}} &= 14{,}000 + \frac{(8.28 \times 10^6)(6000)}{8.28 \times 10^6 + 2.03 \times 10^7} \\ &= 15{,}738 \text{ lbf} \end{aligned}$$

$$\begin{aligned} F_{\text{bolt,max}} &= 14{,}000 + \frac{(8.28 \times 10^6)(12{,}000)}{8.28 \times 10^6 + 2.03 \times 10^7} \\ &= 17{,}477 \text{ lbf} \end{aligned}$$

Stresses in bolt:

$$\begin{aligned} \sigma_{\text{minimum}} &= \frac{15{,}738}{0.6624} \\ &= 23{,}759 \text{ psi} \end{aligned}$$

$$\begin{aligned} \sigma_{\text{maximum}} &= \frac{17{,}477}{0.6624} \\ &= 26{,}384 \text{ psi} \end{aligned}$$

$$\begin{aligned} \sigma_{\text{mean}} &= \left(\tfrac{1}{2}\right)(23{,}759 + 26{,}384) \\ &= 25{,}072 \text{ psi} \end{aligned}$$

$$\begin{aligned} \sigma_{\text{alt}} &= (3.7)\left(\tfrac{1}{2}\right)(26{,}384 - 23{,}759) \\ &= 4856 \text{ psi} \end{aligned}$$

(The thread stress concentration factor is applied to σ_{alt} only.)

S_{ut} is not given, so a Goodman line cannot be drawn. Use the Soderberg line.

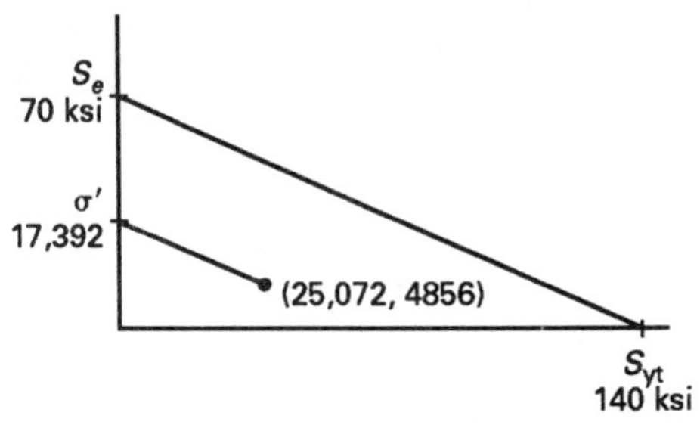

The slope of the Soderberg line is

$$m = -\frac{70}{140} = -0.5$$

The equivalent stress is

$$\begin{aligned} \sigma' &= 4856 + (0.5)(25{,}072) \\ &= 17{,}392 \text{ psi} \end{aligned}$$

The factor of safety is

$$\text{FS} = \frac{70{,}000}{17{,}392} = \boxed{4.02}$$

6. Let gear 1 be the driven gear. Let gear 2 be the pinion.

$$\begin{aligned} N_2 &= 24 \text{ teeth} \\ n_2 &= 1800 \text{ rpm} \\ r_1 + r_2 &= 9.0 \text{ in} \\ \frac{r_1}{r_2} &= 2 \\ r_1 &= 2r_2 \end{aligned}$$

Solving simultaneously,

$$\begin{aligned} r_1 &= 3 \text{ in} \\ r_2 &= 6 \text{ in} \\ d_1 &= \boxed{12 \text{ in}} \\ d_2 &= \boxed{6 \text{ in}} \end{aligned}$$

Since the pinion has 24 teeth, the diametral pitch is

$$P = \frac{24}{6} = 4 \text{ teeth/in}$$

The pitch line velocity is

$$v_r = \frac{\pi(1800)(6)}{12} = 2827 \text{ ft/min}$$

The Barth speed factor is

$$\begin{aligned} k_d &= \frac{1200}{1200 + 2827} = 0.30 \\ Y &= 0.337 \end{aligned}$$

The transmitted force is

$$F = \frac{(50)(33{,}000)}{2827} = 584 \text{ lbf}$$

From the Lewis equation,

$$w = \frac{FP}{Sk_dY} = \frac{(584)(4)}{\left(\frac{38{,}000}{2}\right)(0.30)(0.337)}$$

$$= \boxed{1.22 \text{ in} \quad [\text{say } 1.25 \text{ in}]}$$

Alternatively, a trial and error solution is possible.

7. The force in the gate wires is

$$\frac{800}{2} = 400 \text{ lbf}$$

The torque from the 10-in pulley is

$$T = (400)(5) = 2000 \text{ in-lbf}$$

The force in the motor cable is

$$F = \frac{T}{r} = \frac{2000}{7} = 286 \text{ lbf}$$

Draw the freebody of the shaft.

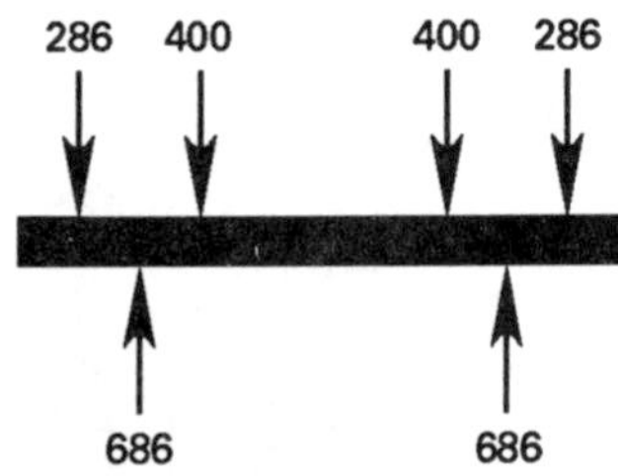

The moment diagram is

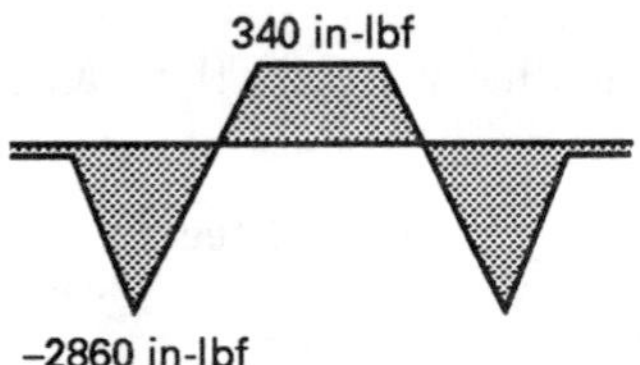

The moment of inertia is

$$I = \tfrac{1}{4}\pi r^4$$

Since the maximum moment is 2860 in-lbf, the maximum stress is

$$\sigma = \frac{Mc}{I} = \frac{2860r}{\frac{1}{4}\pi r^4} = \frac{3642}{r^3}$$

Similarly,

$$J = \tfrac{1}{2}\pi r^4$$

$$\tau = \frac{Tc}{J} = \frac{2000r}{\frac{1}{2}\pi r^4} = \frac{1273}{r^3}$$

From the combined stress equations,

$$\begin{aligned}\sigma_1, \sigma_2 &= \tfrac{1}{2}\sigma \pm \tfrac{1}{2}\sqrt{\sigma^2 + (2\tau)^2} \\ &= \frac{3642}{2r^3} \pm \tfrac{1}{2}\sqrt{\left(\frac{3642}{r^3}\right)^2 + \left[\frac{(2)(1273)}{r^3}\right]^2} \\ &= \frac{1821}{r^3} \pm \frac{2222}{r^3} \\ &= \frac{4043}{r^3}, -\frac{401}{r^3}\end{aligned}$$

Use the distortion energy theory.

$$\sigma_1^2 + \sigma_2^2 - \sigma_1\sigma_2 \le (\sigma_{\text{allowable}})^2$$

$$\left(\frac{4043}{r^3}\right)^2 + \left(\frac{-401}{r^3}\right)^2 - \left(\frac{4043}{r^3}\right)\left(\frac{-401}{r^3}\right) \le \left(\frac{60{,}000}{2}\right)^2$$

$$\frac{4258}{r^3} < 30{,}000$$

$$r = 0.522 \text{ in}$$

Say

$$d = \boxed{1\tfrac{1}{8}\text{-in shaft}}$$

This solution disregards direct shear.

8. Notice that the interference is given, not p. We have to solve for p first.

$$r_i = \frac{8.0 - (2)(0.45)}{2} = 3.55$$

$$r_{\text{interface}} = \frac{8}{2} = 4.00$$

$$r_o \approx \frac{7.99 + (2)(0.30)}{2} = 4.295 \approx 4.30 \text{ in}$$

$$I = |\Delta D_{\text{pipe}}| + |\Delta D_{\text{sleeve}}|$$

$$\Delta D = \left(\frac{D}{E}\right)[\sigma_c - \mu(\sigma_r + \sigma_L)]$$

In this problem, $\sigma_L = 0$.

For a pipe under external pressure,

$$\Delta D_{\text{pipe}} = \left|\left(\frac{D_o}{E}\right)\left[\frac{-(r_o^2 + r_i^2)p}{r_o^2 - r_i^2} - \mu(-p)\right]\right|$$

But $r_i = 3.55$ and $r_o = 4.00$.

$$\Delta D_{\text{pipe}} = \left|\left(\frac{8p}{E}\right)\left[\frac{-[(4)^2 + (3.55)^2]}{(4)^2 - (3.55)^2} + 0.30\right]\right|$$

$$= \frac{64.95p}{E}$$

For a sleeve under internal pressure,

$$\Delta D_{\text{sleeve}} = \left|\left(\frac{D_i}{E}\right)\left[\frac{(r_o^2 + r_i^2)p}{r_o^2 - r_i^2} - \mu(-p)\right]\right|$$

But $r_o = 4.30$ and $r_i = 4.00$.

$$\Delta D_{\text{sleeve}} = \left| \left(\frac{8p}{E}\right) \left[\frac{[(4.30)^2 + (4.00)^2]}{(4.30)^2 - (4.00)^2} + 0.30\right] \right| = \frac{113.2p}{E}$$

The total interference is

$$8.000 - 7.990 = \frac{(64.95 + 113.2)p}{2.9 \times 10^7}$$
$$p = 1628 \text{ psi}$$

Use superposition.

Due to interference:

Pipe, external pressure:

σ_c is maximum at inner face.

$$\sigma_{ci} = \frac{(-2)(4)^2(1628)}{(4)^2 - (3.55)^2} = -15{,}333 \text{ psi}$$

Sleeve, internal pressure:

σ_c is maximum at inner face.

$$\sigma_{ci} = \frac{[(4.30)^2 + (4.00)^2](1628)}{(4.30)^2 - (4.00)^2} = 22{,}550 \text{ psi}$$

Due to internal pressure:

Consider the pipe and sleeve to be homogeneous, one piece with $r_i = 3.55$ and $r_o \approx 4.30$.

Pipe, internal pressure:

$$\sigma_{ci} = \frac{[(4.3)^2 + (3.55)^2](750)}{(4.3)^2 - (3.55)^2} = 3961 \text{ psi}$$

Sleeve, internal pressure:

Evaluate the stress at $r = 4.00$ in.

With $p = 0$,

$$\sigma_c = \frac{r_i^2 p + \dfrac{p_i r_i^2 r_o^2}{r^2}}{r_o^2 - r_i^2} = \frac{(3.55)^2(750) + \dfrac{(750)(3.55)^2(4.30)^2}{(4.00)^2}}{(4.30)^2 - (3.55)^2} = 3461 \text{ psi}$$

The total stresses are

$$\sigma_{c,\text{pipe}} = -15{,}333 + 3961 = \boxed{-11{,}372 \text{ psi compressive}}$$

$$\sigma_{c,\text{sleeve}} = 22{,}555 + 3461 = \boxed{26{,}016 \text{ psi tensile}}$$

9.

$$\omega = \sqrt{\frac{kg}{W}} = 2\pi f$$
$$\frac{kg}{W} = (2\pi f)^2$$

The mass per spring is

$$\frac{2000}{2} = 1000 \text{ lbm}$$
$$k = \left(\frac{W}{g}\right)(2\pi f)^2 = \left(\frac{1000}{32.2}\right)[(2)(\pi)(1.25)]^2 = 1916 \text{ lbf/ft} \quad [160 \text{ lbf/in}]$$

The Wahl factor is

$$W = \frac{(4)(8) - 1}{(4)(8) - 4} + \frac{0.615}{8} = 1.184$$

At solid height, the stress is assumed to be

$$\tau = \frac{60{,}000}{1.75} = 34{,}286 \text{ psi}$$

At solid height, the force in the spring is

$$F = k\delta = (160)(7) = 1120 \text{ lbf}$$
$$d = \sqrt{\frac{8CFW}{\pi\tau}} = \sqrt{\frac{(8)(8)(1120)(1.184)}{\pi(34{,}286)}} = \boxed{0.888 \text{ in} \quad [\text{theoretical-wire size}]}$$

Since $C = 8$,

$$D = (8)(0.888) = \boxed{7.104 \text{ in}}$$
$$n_a = \frac{Gd}{8C^3k} = \frac{(11.5 \times 10^6)(0.888)}{(8)(8)^3(160)} = \boxed{15.6 \text{ coils}}$$

10. This is combined stress.

$$M_{\text{bending}} = 4P \text{ in-lbf}$$

$$I = \tfrac{1}{4}\pi r^4 = \tfrac{1}{4}\pi\left(\frac{1.25}{2}\right)^4 = 0.1198 \text{ in}^4$$

$$\sigma = \frac{Mc}{I} = \frac{(4P)\left(\frac{1.25}{2}\right)}{0.1198} = 20.86P \text{ psi}$$

The torque is

$$T = 8P$$

$$J = \tfrac{1}{2}\pi r^4 = \tfrac{1}{2}\pi\left(\frac{1.25}{2}\right)^4 = 0.2397 \text{ in}^4$$

$$\tau = \frac{Tr}{J} = \frac{(8P)\left(\frac{1.25}{2}\right)}{0.2397} = 20.86P \text{ psi}$$

$$\sigma_y = 0$$

$$\sigma_x = 20.86P$$

$$\tau = 20.86P$$

$$\sigma_1, \sigma_2 = \left(\tfrac{1}{2}\right)(20.86P) \pm \tfrac{1}{2}\sqrt{(20.86P)^2 + [(2)(20.86P)]^2} = 10.43P \pm 23.32P$$

$$\sigma_1 = 33.75P$$

$$\sigma_2 = -12.89P$$

$$\tau_{\text{maximum}} = 23.32P$$

Use the distortion energy theory with the allowable stress in place of S_{yt}.

$$\sigma_1^2 + \sigma_2^2 - \sigma_1\sigma_2 \leq (\sigma_{\text{allowable}})^2$$

$$(33.75P)^2 + (-12.89P)^2 - (33.75P)(-12.89P) \leq (26)^2$$

$$P = 0.623 \text{ kips} \quad \text{[623 lbf]}$$

Since we had no direct control over τ, the allowable shear stress may have been exceeded.

$$\tau_{\text{maximum}} = 23.32P = (23.32)(623) = 14{,}528 > 14{,}000 \text{ psi} \quad \text{[no good]}$$

To avoid a trial and error solution,

$$P_{\text{maximum}} \approx \left(\frac{14{,}000}{14{,}528}\right)(623) = \boxed{600 \text{ lbf}}$$

If the maximum shear stress theory is used with a graphical solution,

$$P \approx 560 \text{ lbf}$$

$$\tau_{\text{maximum}} = (23.32)(560) = 13{,}059 < 14{,}000 \quad \text{[ok]}$$

$$P_{\text{maximum}} = \boxed{560 \text{ lbf}}$$

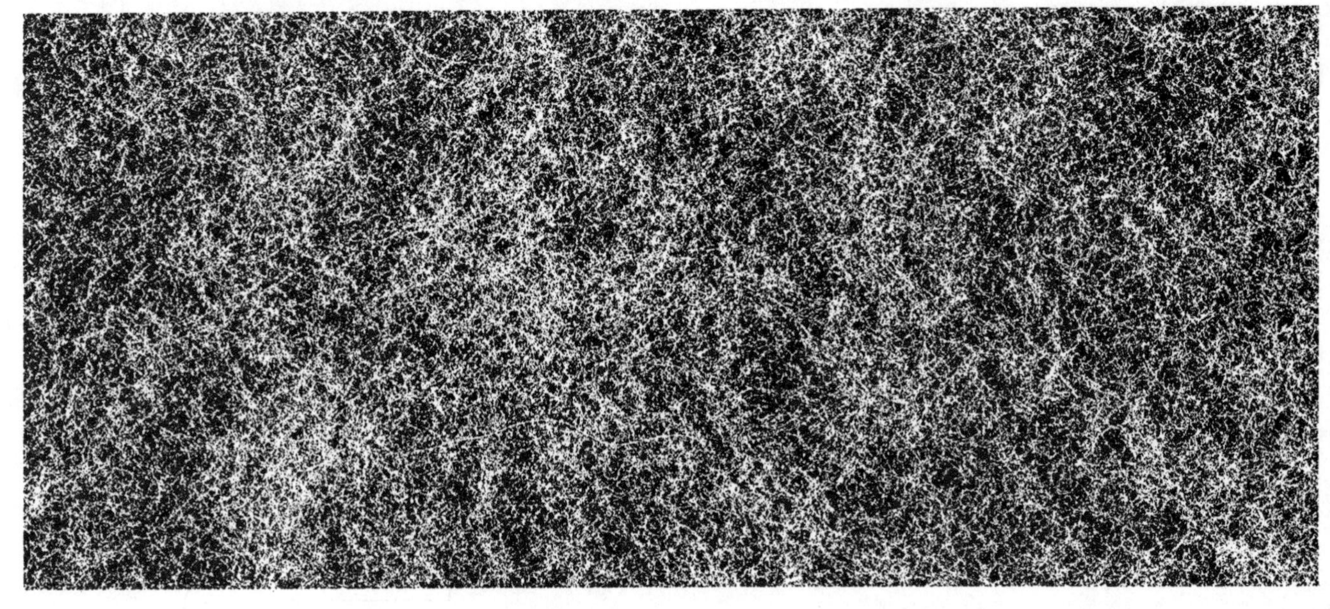

DYNAMICS

1. Use the analogy to traditional planetary gear trains.

Let clockwise be positive.

$$\omega_C = (3000)(0.4) = 1200 \text{ rpm CW}$$

$$\omega_A = \left(\frac{30}{45}\right)(1200) = 800 \text{ rpm CW}$$

$$TV = -1$$

$$\omega_A = \omega_{\text{ring}} = +800 \text{ rpm}$$

$$\omega_{\text{input}} = \omega_{\text{carrier}} = 3000 \text{ rpm}$$

$$\omega_{\text{output}} = \omega_{\text{sun}}$$

$$\omega_{\text{sun}} = TV\omega_r + \omega_C(1 - TV)$$

$$\omega_{\text{sun}} = (-1)(800) + (3000)[1 - (-1)]$$

$$= \boxed{+5200 \text{ rpm CW}}$$

2. Assume we can disregard

- tooth stiffness
- shaft masses

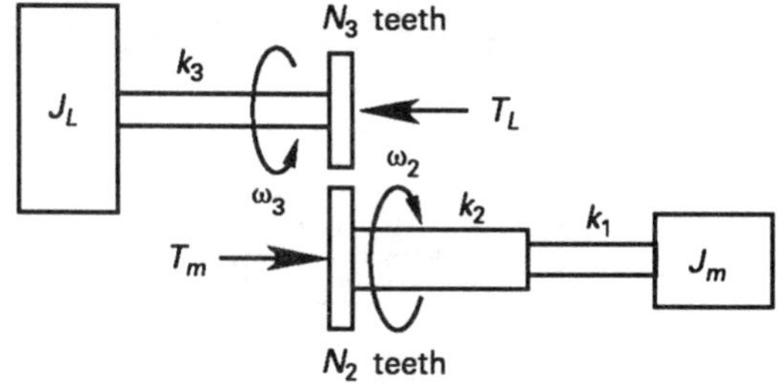

Substitute the following equivalent system.

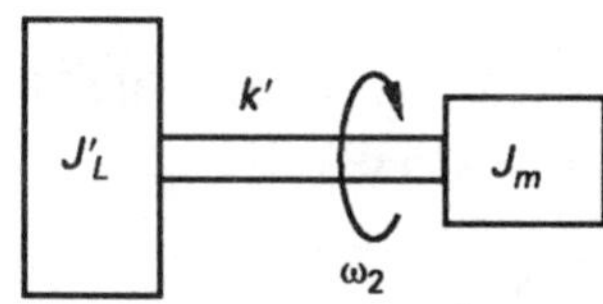

Background and derivations:

To keep the energy transfer across the gear set equal to the equivalent case,

$$\text{K.E.} = \tfrac{1}{2}J_L\omega_3^2 = \tfrac{1}{2}J'_L\omega_2^2$$

Therefore, the equivalent load is

$$J'_L = \left(\frac{\omega_3}{\omega_2}\right)^2 J_L = \left(\frac{N_2}{N_3}\right)^2 J_L$$

Also, the power transmission must be the same.

$$P = T\omega$$

$$P_L = T_L\omega_3$$

$$P' = T'_L\omega_2$$

$$P_L = P'_L$$

Therefore, the equivalent torque is

$$T'_L = \left(\frac{\omega_3}{\omega_2}\right) T_L = \left(\frac{N_2}{N_3}\right) T_L$$

The equivalent rotation is derived from gear train analysis.

$$\theta_2 = \left(\frac{N_3}{N_2}\right)\theta_3$$

The equivalent spring stiffness can be calculated from the foregoing.

$$k = \frac{T}{\theta} \quad [\text{analogous to } k = F/x]$$

$$k' = \frac{T'_L}{\theta_2} = \frac{\left(\frac{N_2}{N_3}\right) T_L}{\left(\frac{N_3}{N_2}\right)\theta_3} = \left(\frac{N_2}{N_3}\right)^2 \left(\frac{T_L}{\theta_3}\right)$$

$$= \left(\frac{N_2}{N_3}\right)^2 k_3 = \left(\frac{D_2}{D_3}\right)^2 k_3$$

Problem solution:

$$k_1 = \frac{T}{\theta} = \frac{GJ}{L} = \frac{G\left(\frac{\pi}{32}\right)d^4}{L}$$

$$= \frac{(11.5 \times 10^6)\left(\frac{\pi}{32}\right)(1)^4}{8}$$

$$= 1.41 \times 10^5 \text{ in-lbf/radian}$$

$$k_2 = \frac{(11.5 \times 10^6)\left(\frac{\pi}{32}\right)(4)^4}{15}$$

$$= 1.93 \times 10^7 \text{ in-lbf/radian}$$

$$k'_3 = \left(\frac{8}{5}\right)^2 \left[\frac{(11.5 \times 10^6)\left(\frac{\pi}{32}\right)(2)^4}{20}\right]$$

$$= 2.31 \times 10^6 \text{ in-lbf/radian}$$

The total equivalent stiffness is

$$\frac{1}{k'} = \frac{1}{k_1} + \frac{1}{k_2} + \frac{1}{k'_3}$$

$$= \frac{1}{1.41 \times 10^5} + \frac{1}{1.93 \times 10^7} + \frac{1}{2.31 \times 10^6}$$

$$k' = 1.32 \times 10^5 \text{ in-lbf/radian}$$

The equivalent load is

$$J_L' = \left(\frac{8}{5}\right)^2 J_L = (2.56)(15) = 38.4$$

The solution to the two-degree of freedom torsion problem is

$$\begin{aligned}\omega_{\text{natural}} &= \sqrt{\frac{k(J_1+J_2)}{J_1 J_2}}\\ &= \sqrt{\frac{(1.32\times 10^5)(25+38.4)}{(25)(38.4)}}\\ &= 93.37 \text{ rad/sec}\\ f_{\text{natural}} &= \frac{93.37}{2\pi} = \boxed{14.9 \text{ Hz}}\end{aligned}$$

3. Assume elastic deflection.

$$\begin{aligned}k_{\text{coils}} &= \frac{Gd^4}{8D^3 n_a} = \frac{(11.5\times 10^6)(0.84)^4}{(8)(6)^3(6)}\\ &= 552.2 \text{ lbf/in each}\end{aligned}$$

For two springs,

$$k_s = (2)(552.2) = 1104.4 \text{ lbf/in}$$

For the two tires,

$$k_{\text{tire}} = (2)(2200) = 4400 \text{ lbf/in}$$

For springs in series, the composite spring constant is

$$\begin{aligned}\frac{1}{k_c} &= \frac{1}{k_1} + \frac{1}{k_2}\\ &= \frac{1}{1104.4} + \frac{1}{4400}\\ k_c &= 882.8 \text{ lbf/in}\end{aligned}$$

(a) This is an inelastic collision, so only momentum is conserved. The velocity of the oven at impact is

$$\begin{aligned}\text{v}_1 &= \sqrt{\text{v}_0^2 + 2as}\\ &= \sqrt{(0)^2 + (2)\left(32.2\ \frac{\text{ft}}{\text{sec}^2}\right)\left(12\ \frac{\text{in}}{\text{ft}}\right)(8 \text{ in})}\\ &= 78.63 \text{ in/sec}\end{aligned}$$

From the conservation of momentum immediately after impact,

$$\begin{aligned}m_1\text{v}_1 + m_2\text{v}_2 &= (m_1+m_2)\text{v}' \quad [\text{v}_2 = 0]\\ \text{v}' &= \frac{m_1\text{v}_1}{m_1+m_2}\\ &= \frac{(800 \text{ lbm})\left(78.63\ \frac{\text{in}}{\text{sec}}\right)}{800 \text{ lbm} + 550 \text{ lbm}}\\ &= 46.6 \text{ in/sec}\end{aligned}$$

The kinetic and potential energy of the oven/truck bed are converted into work as the springs are compressed.

$$\begin{aligned}\Delta E_p + \Delta E_k &= W\\ m' &= m_1 + m_2\\ &= 800 \text{ lbm} + 550 \text{ lbm}\\ &= 1350 \text{ lbm}\end{aligned}$$

$$m'\delta\left(\frac{g}{g_c}\right) + \frac{\frac{1}{2}m'\text{v}^2}{g_c} = \tfrac{1}{2}k\delta^2$$

$$(1350 \text{ lbf})\delta + \frac{\left(\frac{1}{2}\right)(1350 \text{ lbf})\left(46.6\ \frac{\text{in}}{\text{sec}}\right)^2}{\left(32.2\ \frac{\text{lbm-ft}}{\text{lbf-sec}^2}\right)\left(12\ \frac{\text{in}}{\text{ft}}\right)} = \left(\tfrac{1}{2}\right)\left(882.8\ \frac{\text{lbf}}{\text{in}}\right)\delta^2$$

$$\begin{aligned}441.4\delta^2 - 1350\delta - 3793.5 &= 0\\ \delta &= 4.84 \text{ in}\end{aligned}$$

(b) The deflections are inversely proportional to their spring constants.

$$\begin{aligned}\delta_s &= (4.84)\left(\frac{4400}{4400+1104.4}\right)\\ &= 3.87 \text{ in}\end{aligned}$$

(c) At the bottom of the travel, the equivalent force is

$$\begin{aligned}F &= k_t\delta = (882.8)(4.84)\\ &= 4273 \text{ lbf}\end{aligned}$$

The acceleration is

$$\begin{aligned}a &= \frac{F}{m} = \frac{4273}{\frac{800+550}{32.2}}\\ &= \boxed{3.17 \text{ gravities}}\end{aligned}$$

(d)

$$\begin{aligned}S_{yt} &\approx 0.75 S_{ut} = (0.75)(100{,}000)\\ &= 75{,}000 \text{ psi}\\ S_{ys} &= 0.577 S_{yt} = (0.577)(75{,}000)\\ &= 43{,}275 \text{ psi}\\ C &= \frac{6}{0.84} = 7.14\end{aligned}$$

The Wahl factor is

$$\begin{aligned}W &= \frac{(4)(7.14)-1}{(4)(7.14)-4} + \frac{0.615}{7.14}\\ &= 1.21\end{aligned}$$

For one spring,

$$\begin{aligned}F_{\text{maximum}} &= k_s\delta_s = (552.2)(3.87) = 2137 \text{ lbf}\\ \tau &= \frac{8CFW}{\pi d^2} = \frac{(8)(7.14)(2137)(1.21)}{\pi(0.84)^2}\\ &= 66{,}630 \text{ psi}\end{aligned}$$

Since 66,630 > 43,275, this is an

overload condition

4. By definition,

$$TV = \frac{-N_{\text{ring}}}{N_{\text{sun}}} = \frac{-119}{N_{\text{sun}}}$$

Also, $\omega_{\text{sun}}/\omega_{\text{carrier}} = 8$, or $\omega_{\text{carrier}} = \omega_{\text{sun}}/8$.

$$\begin{aligned}\omega_{\text{sun}} &= (TV)\omega_{\text{ring}} + \omega_{\text{carrier}}(1 - TV)\\ &= \left(\frac{-119}{N_{\text{sun}}}\right)(0) + \left(\frac{\omega_{\text{sun}}}{8}\right)\left[1 - \left(\frac{-119}{N_{\text{sun}}}\right)\right]\\ &= \left(\frac{\omega_{\text{sun}}}{8}\right)\left(1 + \frac{119}{N_{\text{sun}}}\right)\\ 8 &= 1 + \frac{119}{N_{\text{sun}}}\\ N_{\text{sun}} &= \boxed{17}\end{aligned}$$

$$D_{\text{sun}} + 2D_{\text{planet}} = D_{\text{ring}}$$

By definition, $P = N/D$, so $D = N/P$.

All teeth have the same diametral pitch, P.

$$\begin{aligned}\frac{N_{\text{sun}}}{P} + \frac{2N_{\text{planet}}}{P} &= \frac{N_{\text{ring}}}{P}\\ N_{\text{sun}} + 2N_{\text{planet}} &= N_{\text{ring}}\\ 17 + 2N_{\text{planet}} &= 119\\ N_{\text{planet}} &= \boxed{51}\end{aligned}$$

5. (a)

$$\omega = 2\pi f = \sqrt{\frac{gk}{m}} \quad [m \text{ in lbm}]$$

$$\begin{aligned}f &= 14 \text{ Hz}\\ m &= 47 \text{ lbm}\\ k &= \frac{f}{\delta} = \frac{12EI}{L^3}\\ (2\pi)(14) &= \sqrt{\frac{(386)(12)(2.9 \times 10^7)I}{(3)^3(47)}}\\ I &= 7.31 \times 10^{-5}\\ &= \frac{bh^3}{12}\\ 7.31 \times 10^{-5} &= (4)\left[\frac{(1)(t)^3}{12}\right]\\ t &= \boxed{0.0603 \text{ in}}\end{aligned}$$

(b) The lowest frequency applied is

$$f_f = \frac{583}{60} = 9.72 \text{ Hz}$$

$$\frac{f_f}{f} > \sqrt{2} \text{ or } f < \frac{f_f}{\sqrt{2}}$$

$$f < \frac{9.72}{\sqrt{2}} = \boxed{6.87 \text{ Hz}}$$

6. Neglect compression braking.

Assume all rotational inertia (momentum) of the propeller is absorbed by the tree.

Neglect the rotational inertia of the shaft.

$$\begin{aligned}E_{k,r} &= \frac{J\omega^2}{2g} = \frac{(100)\left[\frac{(2)\pi(2400)}{60}\right]^2}{(2)(386)}\\ &= 8182 \text{ in-lbf}\end{aligned}$$

The spring constant for a tortional spring is

$$k = \frac{T}{\phi} = \frac{GJ}{L} \quad \left[\frac{\text{in-lbf}}{\text{radian}}\right]$$

$$J = \frac{\pi r^4}{2} = \frac{\pi\left(\frac{2}{2}\right)^4}{2} = 1.57 \text{ in}^4$$

The energy (work) absorbed in twist is

$$W = \tfrac{1}{2}k\phi^2$$

Equating the work and energy,

$$\begin{aligned}E_{k,r} &= W\\ 8182 &= \left(\tfrac{1}{2}\right)\left[\frac{(11.5 \times 10^6)(1.57)}{14}\right]\phi^2\\ \phi &= 0.1126 \text{ radians}\\ \tau &= \frac{Tr}{J} = \frac{\phi GJr}{LJ} = \frac{\phi Gr}{L}\\ &= \frac{(0.1126)(11.5 \times 10^6)\left(\frac{2}{2}\right)}{14}\\ &= \boxed{92{,}493 \text{ psi}}\end{aligned}$$

If this is not in the linear (elastic) region, τ will be limited to S_{ys}.

7. Assume a simply-supported beam loaded at the center. Neglect shaft mass.

(a)

$$f \propto \sqrt{\frac{1}{\delta_{st}}}$$

$$\delta_{st} \propto L^3$$

Therefore,

$$f \propto \sqrt{\left(\frac{1}{L}\right)^3}$$

$$\frac{f_{\text{new}}}{f_{\text{old}}} = \sqrt{\left(\frac{1}{1.15}\right)^3} = 0.81$$

$$\boxed{81\% \text{ of original speed}}$$

(b)

$$f \propto \sqrt{\frac{1}{\delta_{st}}}$$

$$\delta_{st} \propto \frac{1}{I} = \frac{1}{D^4}$$

Therefore,

$$f \propto \sqrt{D^4} = D^2$$

$$\frac{f_{\text{new}}}{f_{\text{old}}} = (1.15)^2 = 1.32$$

$$\boxed{132\% \text{ of original speed}}$$

(c)

$$m_{\text{rotor}} = \rho A t \propto \frac{\pi}{4} D^2 t$$

$$f \propto \sqrt{\frac{1}{\delta_{st}}}$$

$$\delta_{st} \propto F \propto m \propto D^2$$

Therefore,

$$f \propto \sqrt{\frac{1}{D^2}} = \frac{1}{D}$$

$$\frac{f_{\text{new}}}{f_{\text{old}}} = \frac{1}{1.15} = 0.87$$

$$\boxed{87\% \text{ of original speed}}$$

(d)

$$m_{\text{rotor}} = \rho A t \propto \frac{\pi}{4} D^2 t$$

$$f \propto \sqrt{\frac{1}{\delta_{st}}}$$

$$\delta_{st} \propto F \propto m \propto t$$

Therefore,

$$f \propto \sqrt{\frac{1}{t}}$$

$$\frac{f_{\text{new}}}{f_{\text{old}}} = \sqrt{\frac{1}{1.15}} = 0.93$$

$$\boxed{93\% \text{ of original speed}}$$

8. Kinetic energy is not a factor at the terminal velocity since it is constant.

The drop in potential energy equals the rotational energy increase of the rollers.

(a) Block falling:

Let the block slide some distance, say 15 in, down the conveyor. The vertical drop is

$$\Delta z = (15)\sin 35^\circ = 8.6 \text{ in}$$

The decrease in potential energy is (m in lbm)

$$\Delta E_p = w\Delta z = \left(\frac{gm}{g_c}\right)\Delta z = (40)(8.6) = 344 \text{ in-lbf}$$

At the terminal velocity, v_t in/sec, the time required to travel 15 in down the conveyor is

$$t = \frac{s}{\mathrm{v}_t} = \frac{15}{\mathrm{v}_t}$$

This is the same time as required to drop 8.6 in vertically.

The rate of energy change (power) is

$$P_{\text{block}} = \frac{\Delta E_p}{t} = \frac{344\mathrm{v}_t}{15} = 22.9\mathrm{v}_t \text{ in-lbf/sec}$$

Rollers:

The roller circumference is

$$\pi d = (\pi)(2) = 6.283 \text{ in}$$

The tangential velocity of a roller is

$$\mathrm{v}_t = \left(n\ \frac{\text{rev}}{\text{sec}}\right)(\text{circumference}) = (n)(\pi d) = 6.283n$$

With no slipping, each roller in contact with the box must have a tangential velocity equal to the final block velocity.

In being brought up to rotational speed, a roller gains rotational kinetic energy equal to

$$\Delta E_{k,r} = \frac{I\omega^2}{2g_c}$$

$$I = \tfrac{1}{2}m_r r^2 = \left(\tfrac{1}{2}\right)(10)\left(\frac{2}{2}\right)^2 = 5 \text{ lbm-in}^2$$

$$\omega = 2\pi n \text{ radians/sec}$$

$$g_c = 386 \text{ in-lbm/lbf-sec}^2$$

Therefore, the rotational kinetic energy is

$$\Delta E_{k,r} = \frac{(5)(2\pi n)^2}{(2)(386)} = 0.256n^2 \text{ per roller}$$

In traveling 15 in down the ramp, the number of rollers achieving this rotational kinetic energy is

$$\frac{15}{3} = 5 \text{ rollers}$$

The total rotational kinetic energy is

$$\Delta E_{k,r} = (5)(0.256)n^2 = 1.28n^2$$

However, it is known that

$$n = \frac{\text{v}_t}{6.283}$$

Therefore,

$$\Delta E_{k,r} = (1.28)\left(\frac{\text{v}_t}{6.283}\right)^2 = 0.0324\text{v}_t^2$$

Since it takes $t = 15/\text{v}_t$ sec to start 5 rollers, the power is

$$P_{\text{rollers}} = \frac{\Delta E_{k,r}}{t} = \frac{0.0324\text{v}_t^2}{\frac{15}{\text{v}_t}} = 0.00216\text{v}_t^3$$

Equating power:

$$\begin{aligned} \text{power in} &= \text{power out} \\ P_{\text{rollers}} &= P_{\text{block}} \\ 22.9\text{v}_t &= 0.00216\text{v}_t^3 \\ \text{v}_t &= \boxed{103 \text{ in/sec} \quad [5.85 \text{ mph}]} \end{aligned}$$

(b) The decelerating work is done by the spring and friction.

Friction:

$$\begin{aligned} F_f &= fN = (0.15)(40) = 6 \text{ lbf} \quad [\text{constant}] \\ W_f &= F_f\delta = 6\delta \end{aligned}$$

Spring:

$$\begin{aligned} F_{s,\text{maximum}} &= k\delta = 80\delta \\ W_{\text{average}} &= \tfrac{1}{2}F_{s,\text{maximum}}\delta = 40\delta^2 \end{aligned}$$

From conservation of energy,

$$E_k = W_f + W_s$$

The incoming kinetic energy is

$$E_k = \frac{m\text{v}^2}{2g_c} = \frac{(40)(80)^2}{(2)(386)} = 331.6 \text{ in-lbf}$$

$$331.6 = 6\delta + 40\delta^2$$

Solving,

$$\delta = 2.81 \text{ in}$$

The maximum force on the block is

$$F_{\text{maximum}} = 6 + (80)(2.81) = 230.8$$

The acceleration is

$$\begin{aligned} a_{\text{maximum}} &= \frac{F_{\text{maximum}}}{\frac{m}{g_c}} = \frac{230.8}{\frac{40}{32.2}} \\ &= 185.8 \text{ ft/sec}^2 \\ &= \frac{185.8}{32.2} = \boxed{5.77 \text{ g}} \end{aligned}$$

(c) If $a_{\text{maximum}} = (2)(32.2) = 64.4 \text{ ft/sec}^2$, then

$$\begin{aligned} F_{\text{maximum}} &= \left(\frac{m}{g_c}\right)a_{\text{maximum}} = \left(\frac{40}{32.2}\right)(64.4) \\ &= 80 \text{ lbf} \\ F_{\text{spring,maximum}} &= F_{\text{maximum}} - F_f = 80 - 6 \\ &= 74 \text{ lbf} \\ E_k &= W_f + W_s \\ &= F_f\delta + \tfrac{1}{2}F_{s,\text{maximum}}\delta \\ 331.6 &= 6\delta + \left(\tfrac{1}{2}\right)(74)\delta \\ \delta &= 7.71 \text{ in} \\ k &= \frac{F}{\delta} = \frac{74}{7.71} \\ &= \boxed{9.60 \text{ lbf/in}} \end{aligned}$$

9. The included angle (wrap angle) for the smaller pulley is

$$(2)(60) = 120° = 2.09 \text{ radians}$$

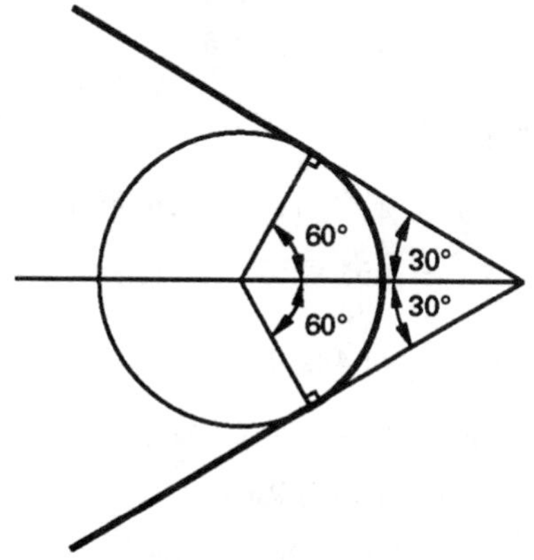

The smaller pulley has a smaller contact angle and limits its torque transmission.

$$\frac{T_1 - F_c}{T_2 - F_c} = e^{(0.35)(2.09)} = 2.08$$
$$T_1 = 2.08T_2$$

(This analysis neglects centrifugal force, F_c, since the belt mass is unknown.)

For the smaller pulley with a radius of $r = 3/2$,

$$10 = (T_1 - T_2)\left(\frac{\frac{3}{2}}{12}\right) \quad [1]$$

$$T_1 - T_2 = 80 \quad [2]$$

Solving these two equations simultaneously,

$$T_1 = \boxed{154.1 \text{ lbf}}$$

$$T_2 = \boxed{74.1 \text{ lbf}}$$

The x-component of the tension sum is

$$(74.1 + 154.1)\cos 30° = 197.6 \text{ lbf}$$

(Both tensions are in the same direction.)

The spring deflection is

$$\delta = \frac{F}{k} = \frac{197.6}{65} = \boxed{3.0 \text{ in}}$$

10. The inside diameter is

$$0.20 \text{ in} - (2)(0.025 \text{ in}) = 0.15 \text{ in}$$

The cross-sectional area is

$$A = \frac{\left(\frac{\pi}{4}\right)\left[(0.20)^2 - (0.15)^2\right]}{144} = 9.54 \times 10^{-5} \text{ ft}^2$$

The volume per foot is

$$V = AL = (9.54 \times 10^{-5})(1) = 9.54 \times 10^{-5} \text{ ft}^3/\text{ft}$$

Steel has a density of 489 lbm/ft^3, so the mass per foot is

$$m = \rho V = (489)(9.54 \times 10^{-5}) = 0.04665 \text{ lbm/ft} = 0.003888 \text{ lbm/in}$$

The area moment of inertia is

$$I = \frac{\pi}{4}r^4 = \frac{\pi}{64}d^4 = \left(\frac{\pi}{64}\right)\left[(0.20)^4 - (0.15)^4\right] = 5.37 \times 10^{-5} \text{ in}^4$$

The modal frequencies are

$$f_n = \frac{K_n}{2\pi}\sqrt{\frac{EIg}{wL^4}}$$

w is the distributed weight per unit length.

K_n is a constant that depends on the mode.

		node positions (x/L)		
mode	K_n	1st	2nd	3rd
1	3.52	0	–	–
2	22.0	0	0.783	–
3	61.7	0	0.504	0.868

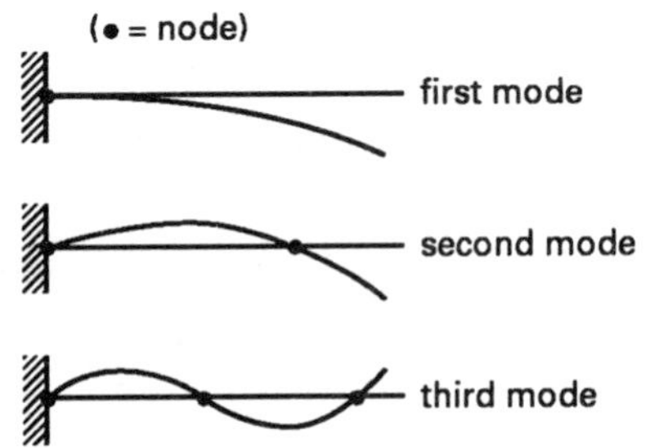

The second resonant frequency is

$$f_2 = \frac{22.0}{2\pi}\sqrt{\frac{(2.9 \times 10^7)(5.37 \times 10^{-5})(386)}{(0.003888)(35)^4}} = \boxed{35.5 \text{ Hz}}$$

11.

(a)

$$W = \text{work}$$
$$W_{\text{deflection}} = \Delta E_{\text{potential}} = wh \quad [w \text{ in pounds}] = mgh \quad [m \text{ in slugs}]$$

This energy must be absorbed by the spring action of the joist.

For a spring, $W = \frac{1}{2}kx^2$ in the elastic region.

$$k = \text{spring constant} = \frac{F}{x}$$

$$k = \frac{F}{y_{\text{maximum}}} = \frac{48EI}{L^3} = \frac{(48)\left(2.9 \times 10^7 \frac{\text{lbf}}{\text{in}^2}\right)(53.4 \text{ in}^4)}{\left[(24 \text{ ft})\left(12 \frac{\text{in}}{\text{ft}}\right)\right]^3} = 3111.7 \text{ lbf/in}$$

Let x be the joist deflection.

$$\begin{aligned}\Delta E_p &= W_{\text{deflection}}\\ mgh &= \tfrac{1}{2}kx^2\\ (1300)[(2)(12)+x] &= \left(\tfrac{1}{2}\right)(3111.7)x^2\\ 31{,}200+1300x &= 1555.9x^2\end{aligned}$$

In quadratic form, this is

$$x^2 - 0.836x - 20.05 = 0$$

Completing the square,

$$\begin{aligned}x^2 - 0.836x &= 20.05\\ (x-0.418)^2 &= 20.05 + (0.418)^2\\ x - 0.418 &= \pm\sqrt{20.22}\\ &= \pm 4.497\\ x &= 4.916 \text{ in}\end{aligned}$$

Now, calculate the moment acting on the beam from the deflection.

The moment at maximum deflection is

$$\begin{aligned}y_{\text{maximum}} &= \frac{FL^3}{48EI} = \frac{(4)\left(\frac{1}{4}FL\right)L^2}{48EI}\\ &= \frac{4M_{\text{maximum}}L^2}{48EI}\\ M_{\text{maximum}} &= \frac{12y_{\text{maximum}}EI}{L^2}\\ &= \frac{(12)(4.916)(2.9\times10^7)(53.4)}{[(24)(12)]^2}\\ &= 1.10\times10^6 \text{ in-lbf}\end{aligned}$$

The stress is

$$\begin{aligned}\sigma &= \frac{Mc}{I} = \frac{M}{S} = \frac{1.10\times10^6}{16.7}\\ &= 65{,}900 \text{ psi}\end{aligned}$$

Since $\sigma > S_{yt}$, yielding will occur and this high stress will never be seen.

This also means that the use of the elastic equation $F = kx$ was inappropriate, and x is wrong.

An impact factor may have some bearing on this problem, but none is specified.

Also, some reduction in potential energy is possible, but only if the joist mass is included in the analysis.

Joists will yield.

(b) The beam (joist) will yield, and the resisting force will be maximum during yielding.

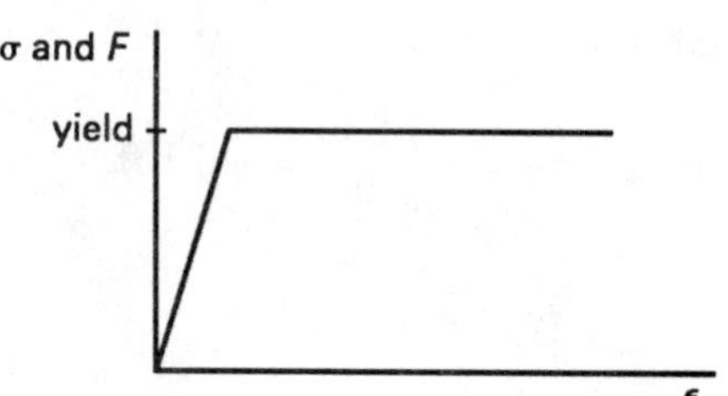

$$\begin{aligned}M_{\text{maximum}} &= \tfrac{1}{4}F_{\text{maximum}}L\\ \sigma &= \frac{Mc}{I} = \frac{\frac{1}{4}F_{\text{maximum}}L}{S}\\ S &= \frac{I}{c} = \text{section modulus}\end{aligned}$$

At yield, $\sigma = S_{yt}$, so

$$\begin{aligned}F_{\text{maximum}} &= \frac{4S_{yt}S}{L} = \frac{(4)(36{,}000)(16.7)}{(24)(12)}\\ &= 8350 \text{ lbf}\end{aligned}$$

From $F = ma$,

$$\begin{aligned}a_{\text{maximum}} &= \frac{F_{\text{maximum}}}{m} \quad [m \text{ in slugs}]\\ &= \frac{8350}{\frac{1300}{32.2}} = 206.8 \text{ ft/sec}^2\\ &= \frac{206.8\ \frac{\text{ft}}{\text{sec}^2}}{32.2\ \frac{\text{ft}}{\text{sec}^2\text{-gravity}}} = 6.4 \text{ gravities}\end{aligned}$$

$6.4 < 8.0$, no

12. There are two ways of approaching this problem.

(A) A system encounters a step forcing function.

(B) A system is released from static compression.

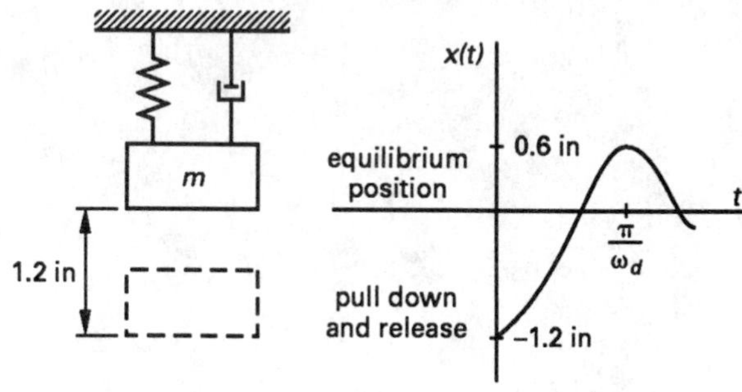

Since there is overshoot, the system is underdamped.

The general solution is

$$x(t) = e^{-nt}(C_1 \cos\omega_d t + C_2 \sin\omega_d t)$$

The initial conditions are

$$x_0 = \frac{-1.2 \text{ in}}{12 \frac{\text{in}}{\text{ft}}} = -0.1 \text{ ft}$$

$$v_0 = 0$$

Substituting $t = 0$ into $x(t)$,

$$C_1 = -0.1$$
$$x(t) = e^{-nt}[(-0.1)\cos\omega_d t + C_2 \sin\omega_d t]$$

(c) The natural frequency is

$$\omega = \sqrt{\frac{kg}{w}} = \sqrt{\frac{(90)(32.2)(12)}{1200}}$$
$$= \boxed{5.38 \text{ rad/sec} \quad [0.856 \text{ Hz}]}$$

(a) At $t = \pi/\omega_d$,

$$x(t) = \frac{0.6 \text{ in}}{12 \frac{\text{in}}{\text{ft}}} = 0.05 \text{ ft}$$

Substituting into $x(t)$ and noting that $\cos(\pi) = -1$ and $\sin(\pi) = 0$,

$$0.05 = (-0.1)e^{\frac{-n\pi}{\omega_d}} \cos\pi$$
$$= (-0.1)e^{\frac{-n\pi}{\omega_d}}(-1)$$
$$0.05 = (0.1)e^{\frac{-n\pi}{\omega_d}}$$

Dividing both sides by 0.1,

$$0.5 = e^{\frac{-n\pi}{\omega_d}}$$

Taking the natural log of both sides,

$$-0.693 = \frac{-n\pi}{\omega_d}$$
$$\omega_d = \sqrt{\omega^2 - n^2}$$
$$-0.693 = \frac{-n\pi}{\sqrt{\omega^2 - n^2}}$$

Solving,

$$n^2 = \frac{\omega^2}{21.54}$$

From part (a), $\omega = 5.38$ rad/sec, so

$$n = \sqrt{\frac{(5.38)^2}{21.54}} = 1.16$$
$$C = \frac{2mn}{g_c} = \frac{(2)(1200)(1.16)}{32.2}$$
$$= \boxed{86.5 \text{ lbf-sec/ft}}$$

(b) The waveform envelope decreases according to the scaling factor e^{-nt}.

For a reduction of the amplitude to 20% of its original value,

$$\frac{x}{x_{\text{maximum}}} = 0.2 = e^{-1.16t}$$

Taking logs of both sides,

$$\ln(0.2) = -1.16t$$
$$t = \boxed{1.39 \text{ sec}}$$

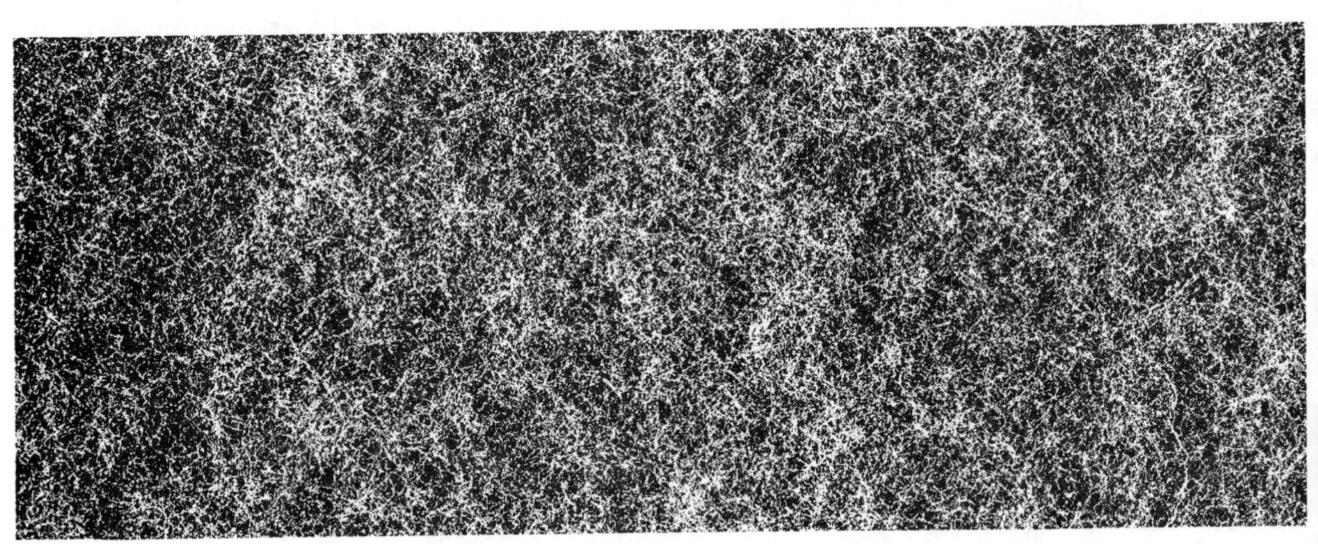

CONTROLS AND SYSTEMS

1. (a)

$$h_{\text{steam}} = 1149.7 \text{ BTU/lbm}$$
$$h_{60^\circ\text{F water}} = 28.06 \text{ BTU/lbm}$$
$$h_{180^\circ\text{F water}} = 147.92 \text{ BTU/lbm}$$

The steady-state energy balance is

$$\dot{m}_{\text{steam}}h_{\text{steam}} + \dot{m}_{60^\circ\text{F}}h_{60^\circ\text{F}} = \dot{m}_{180^\circ\text{F}}h_{180^\circ\text{F}}$$
$$\dot{m}_{\text{steam}}(1149.7) + (30 - \dot{m}_{\text{steam}})(28.06) = (30)(147.92)$$
$$\dot{m}_{\text{steam}} = 3.21 \text{ lbm/min}$$
$$\dot{m}_{60^\circ\text{F water}} = 30 - 3.21 = 26.79 \text{ lbm/min}$$

Since the valve is either fully open or fully closed, the time must be chosen to achieve this flow rate.

$$t = \left(\frac{3.21}{15}\right)\left(60\ \frac{\text{sec}}{\text{min}}\right) = \boxed{12.84 \text{ sec open/min}}$$

(b) This is a changing-volume fluid mixture problem. The "solute" is energy.

Assume

$$t = 0 \text{ when valve sticks}$$
$$c_p = 1 \text{ BTU/lbm-}^\circ\text{F for water}$$

The entering energy is

$$U_{\text{in}} = \dot{m}_{\text{in}}h_{\text{in}} = (15)(1149.7) + (26.79)(28.06) = 18{,}000 \text{ BTU/min}$$

The leaving energy is

$$U_{\text{out}} = \dot{m}_{\text{out}}h_{\text{out}} = 30h_{\text{out}}$$

The rate of change of energy in the tank is

$$\frac{dU(t)}{dt} = U_{\text{in}} - U_{\text{out}} = 18{,}000 - 30h_{\text{out}}$$

Since m_{final} is not known, U_{final} is also not known. A better equation would be in terms of $T(t)$, the temperature at time t.

Convert the $dU(t)/dt$ equation to $dT(t)/dt$.

$$h_{\text{out}} = c_p(T - 32) \approx T - 32$$

$dU(t)/dt$ is found from $U(t)$.

$$U(t) = m(t)h(t)$$
$$m(t) = 600 + (15 + 28.06 - 30)t = 600 + 13.06t$$
$$h(t) = c_p[T(t) - 32] \approx T(t) - 32$$
$$U(t) = (600 + 13.06t)[T(t) - 32] = 600T(t) - 19{,}200 + 13.06tT(t) - 418t$$
$$U'(t) = \frac{dU(t)}{dt} = 600T'(t) + (13.06)[tT'(t) + T(t)] - 418$$
$$= 600T'(t) + 13.06tT'(t) + 13.06T(t) - 418$$
$$= T'(t)[600 + 13.06t] + 13.06t - 418$$

Combining into the differential equation,

$$U'(t) = U_{\text{in}} - U_{\text{out}}$$
$$T'(t)[600 + 13.06t] + 13.06T(t) - 418 = 18{,}000 - 30[T(t) - 32]$$

The differential equation is

$$T'(t)(600 + 13.06t) + 43.06T(t) - 19{,}378 = 0$$

This is a first-order, linear equation. Putting it in standard form,

$$T'(t) + \left(\frac{43.06}{600 + 13.06t}\right)T(t) = \frac{19{,}378}{600 + 13.06t}$$

The integrating factor is

$$\mu(t) = \exp\left(\int \frac{43.06}{600 + 13.06t}dt\right)$$
$$= \exp\left(43.06\int \frac{dt}{600 + 13.06t}\right)$$
$$= \exp\left(43.06\left[\frac{1}{13.06}\ln(13.06t + 600)\right]\right)$$
$$= \exp[3.3\ln(13.06t + 600)]$$
$$= e^{3.3\ln(13.06t+600)}$$
$$= \left[e^{\ln(13.06t+600)}\right]^{3.3}$$
$$= [\ln(13.06t + 600)]^{3.3}$$

$$T(t) = \frac{1}{[\ln(13.06t + 600)]^{3.3}} \times \int \frac{[\ln(13.06t+600)]^{3.3}(19{,}378)}{600 + 13.06t} dt$$
$$= \left(\frac{1}{[\ln(13.06t+600)]^{3.3}}\right) \times \left(\left[\frac{19{,}378}{(4.3)(13.06)}\right][\ln(13.06t+600)]^{4.3} + C\right)$$
$$= 345\ln(13.06t+600) + \frac{C}{[\ln(13.06t+600)]^{3.3}}$$

Solve for C. $T = 180°\text{F}$ at t = 0.

$$180 = 345\ln(600) + \frac{C}{[\ln(600)]^{3.3}}$$
$$C = -925{,}864$$

The temperature as a function of time is

$$T(t) = 345\ln(13.06t+600) - \frac{925{,}864}{[\ln(13.06t+600)]^{3.3}}$$

Solve for t when $T = 190$ by trial and error.

t	$T(t)$
0	180
1	209.8
0.4	192.0
0.33	189.9

$t \approx$ 0.33 min

(c) The proportional band is the range of temperatures necessary to move the control valve from fully closed to fully open. It is given in the problem as

PB = 12°F

(d) set point = 180°F

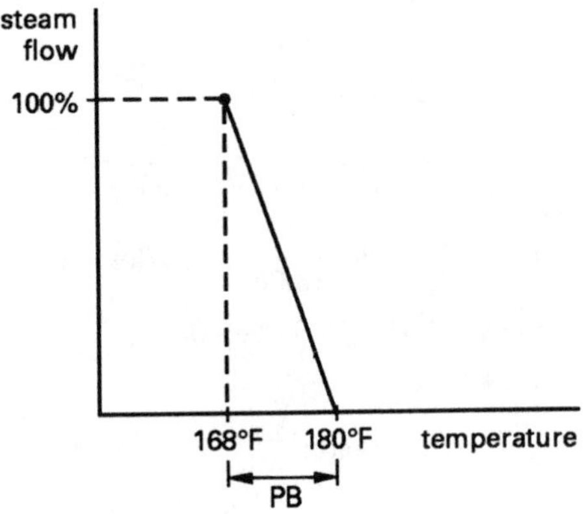

2. This is a linear programming problem.

N_t = number of turbo cycles produced
N_m = number of mopeds produced

The objective (function) is to maximize profit.

$$\text{maximum } Z = 500N_t + 300N_m$$

The constraints are

$$6N_t + 2N_m \le 1800 \quad [1]$$
$$4N_t + 2N_m \le 1400 \quad [2]$$
$$6N_t + 8N_m \le 4400 \quad [3]$$
$$N_t \ge 0$$
$$N_m \ge 0$$

Since this is a two-dimensional problem, it can be solved graphically.

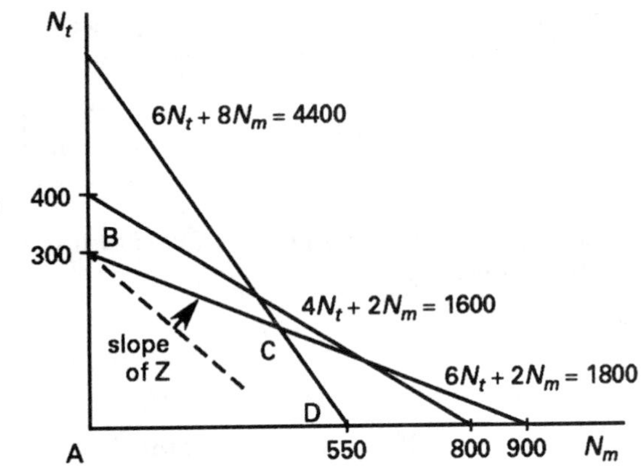

At point C, the first and third constraints intersect. That point is

$$\begin{aligned} -1 \times [1]: &\quad -6N_t - 2N_m = -1800 \\ [3]: &\quad 6N_t + 8N_m = 4400 \\ \hline &\quad 6N_m = 2600 \end{aligned}$$

$N_m = 433.33$ [say 433]
$N_t = 155.6$ [say 155]

Although it is clear that point C is the maximum profit point (since the Z line touches it last as it moves out), we can calculate profit at all extreme points to prove it.

Point	N_t	N_m	Z
A	0	0	0
B	300	0	150,000
C	155	433	207,400
D	0	550	165,000

The strategy that maximizes profit is

$N_t = 155$
$N_m = 433$

3. This is negative feedback.

(a) The transfer function is

$$T(s) = \frac{G(s)}{1+G(s)H(s)}$$

$$= \frac{\frac{1}{s^2+2s+1}}{1+\left(\frac{1}{s^2+2s+1}\right)\left(\frac{K}{s+3}\right)}$$

$$= \frac{\frac{1}{s^2+2s+1}}{\frac{(s^2+2s+1)(s+3)+K}{(s^2+2s+1)(s+3)}}$$

$$= \frac{s+3}{(s^2+2s+1)(s+3)+K}$$

$$= \boxed{\frac{s+3}{(s+1)(s+1)(s+3)+K}}$$

(b) Use the Routh stability criterion. The expanded characteristic equation is

$$(s+1)(s+1)(s+3)+K = s^3+2s^2+s+3s^2 + 6s+3+K = s^3+5s^2+7s+3+K$$

The Routh table is

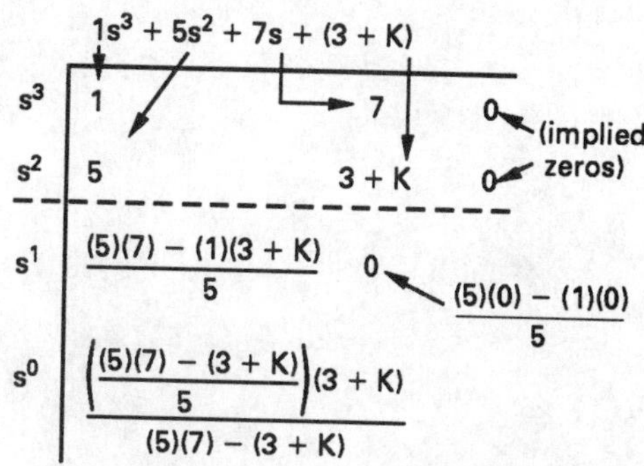

Simplifying,

s^3	1	7	0
s^2	5	3 + K	0
s^1	$\frac{32-K}{5}$	0	
s^0	3 + K	0	

To be stable, the first column must not change signs. The feedback loop will be stable if

$$\boxed{32 > K > -3}$$